Psychiatry Under the Influence

Other books by Robert Whitaker

Mad in America: Bad Science, Bad Medicine, and the Enduring Mistreatment of the Mentally Ill (2002)
The Mapmaker's Wife: A True Tale of Love, Murder, and Survival in the Amazon (2004)
On the Laps of Gods: The Red Summer of 1919 and the Struggle for Justice that Remade a Nation (2008)
Anatomy of an Epidemic: Magic Bullets, Psychiatric Drugs, and the Astonishing Rise of Mental Illness in America (2010)

Other books by Lisa Cosgrove

The Ethics of Pharmaceutical Industry Influence in Medicine (coauthored with O. S. Haque, J. De Freitas, H. J. Bursztajn, A. Gopal, R. Paul, I. Shuv-Ami, and S. Wolfman, S., 2013).
Bias in Psychiatric Diagnosis (coedited with P. J. Caplan, 2004).

Psychiatry Under the Influence

Institutional Corruption, Social Injury, and Prescriptions for Reform

Robert Whitaker
and
Lisa Cosgrove

First published in 2015 by
PALGRAVE MACMILLAN®
in the United States—a division of St. Martin's Press LLC,
175 Fifth Avenue, New York, NY 10010.

Where this book is distributed in the UK, Europe and the rest of the world,
this is by Palgrave Macmillan, a division of Macmillan Publishers Limited,
registered in England, company number 785998, of Houndmills,
Basingstoke, Hampshire RG21 6XS.

Palgrave Macmillan is the global academic imprint of the above companies
and has companies and representatives throughout the world.

Palgrave® and Macmillan® are registered trademarks in the United States,
the United Kingdom, Europe and other countries.

ISBN: 978–1–137–50694–8 (hc)
ISBN: 978–1–137–50692–4 (pbk)

Library of Congress Cataloging-in-Publication Data

Whitaker, Robert, author.
    Psychiatry under the influence : institutional corruption, social injury and
prescriptions for reform / Robert Whitaker and Lisa Cosgrove.
        p. ; cm.
    ISBN 978–1–137–50692–4 (paperpack)—
    ISBN 978–1–137–50694–8 (hardcover)
        I. Cosgrove, Lisa, author. II. Title. [DNLM: 1. American Psychiatric Association.
2. Psychiatry—history—United States. 3. Societies, Medical—history—United
States. 4. Conflict of Interest—United States. 5. Ethics, Institutional—history—
United States. 6. History, 20th Century—United States. 7. Psychiatry—ethics—
United States. WM 1]
    RC454.4
    616.89—dc23   2014043052

A catalogue record of the book is available from the British Library.

Design by Newgen Knowledge Works (P) Ltd., Chennai, India.

First edition: April 2015

D 10 9 8 7 6 5 4 3

To the Edmond J. Safra Research Lab on Institutional Corruption at Harvard University, which made this book possible

CONTENTS

FIGURES AND TABLES

Figures

Tables

FOREWORD

LAWRENCE LESSIG*

In 2010, Harvard's Edmond J. Safra Center for Ethics launched a "Lab" to study "institutional corruption." Our aim was not to target bad souls doing bad things. Instead, the corruption that we were interested in was more ordinary, or regular. It was the product of a set of influences, within economies of influence, that weaken the effectiveness of the particular institution, especially by weakening public trust.

I was drawn to this conception of corruption through my own reflections on the dysfunction of Congress. Congress is not filled with criminals. Yet it seems clear enough that the institution has allowed the influence of campaign funding to weaken its effectiveness, and certainly weaken its public trust. My own work developed the analysis of Congress as an example of "institutional corruption." Indeed, in my view, the example was paradigmatic.

But early in the life of the Lab, Professor Lisa Cosgrove described a similar dynamic within the field of psychiatry. As she explained it in presentations and in her writings, psychiatry too had allowed itself to be affected by an influence that had corrupted its core mission—to help patients. The dynamic of that corruption was different from the dynamic in Congress. But as she believed, the story could be understood in similar terms. The field had been vulnerable, she described, to the commercial influence of pharmaceutical companies. And the story of that influence, she argued, could be another paradigmatic case for "institutional corruption."

This book proves that she was right. In an incredibly compelling and convincing account, Lisa Cosgrove and Robert Whitaker show just how and why psychiatry has been corrupted by the influence of the pharmaceutical industry. And in so doing, they may well have described an even clearer case of institutional corruption than Congress. The actors in this story are not evil, even if there are a few one would be tempted to use that

* Roy L. Furman Professor of Law and Leadership, Harvard Law School; Director, Edmond J. Safra Center for Ethics, Harvard University.

term to describe. They are instead responding to understandable, if pedestrian, pressures. But through a slow, if pronounced, progress, the very aim of the field becomes linked to an influence that conflicts with the core purpose of any institution within medicine—to help the patient. And the consequence of that conflict is a practice and an industry that consumes 6.3 percent of health care costs, with, except for the most extreme cases, very little demonstrable benefit to the patient.

No doubt this book will attract incredible scrutiny. And indeed, given the interests at stake, it is certain to be attacked, as will its authors. But Cosgrove's academic and professional experience, combined with Whitaker's powerful investigative reporting, provide an account that will astonish the fair reader. And when the arguments they make are accepted, as they will be, they will induce a fundamental change in how we approach the field of mental health. Whether or not there was an excuse for what broad swaths of psychiatry became, there can be little excuse for leaving it untouched now.

ACKNOWLEDGMENTS

As we note in our introduction, this book arose from our time as fellows at the Edmond J. Safra Research Lab at Harvard University, which was established to further the study of institutional corruption. We are grateful to Mrs. Lily Safra for her generous funding of this extraordinary Lab, which provided us with a rare opportunity to focus our energies on this topic. Lawrence Lessig inspired us to tackle this project, and under his leadership, the Lab has developed a framework for investigating institutional corruption that guided our thinking, and our construction of this book, at every step. The framework enabled us to "see" this story of corruption in a way that was new for both of us.

Of the two of us, Lisa has been associated with the Lab since its inception in 2010, and during that time, she has enjoyed the friendship of so many there. Together, we have numerous people at the Lab to thank for their encouragement, feedback, and support: Sheila Kaplan, Stephanie Dant, Neeru Paharia, Mark Somos, William English, Malcolm Salter, Jonathan Marks, Heidi Carrell, Katy Evans Pritchard, Mark Rodwin, Garry Gray, and Donald Light. We would also like to thank Emily Wheeler, Shannon Peters, and Brianna Goodale for their considerable editorial assistance, and similarly thank Scott Greenspan, Jenesse Kaitz, and Erin Murtagh for their contributions to this project. We are grateful to Cynthia Frawley for her construction of the graphics, and to Lisa Rivero for creating the index.

During these past years, Lisa has been fortunate to call Harold Bursztajn and Shelly Krimsky her mentors, and to consider them as friends as well. She also was fortunate to collaborate with Allen Shaughnessy on various studies that are cited in this book. She wants to thank Doreen Hiltz, Kelly Atkinson, Lisa Hurley, Dan Torres, and Sharon Lamb for their friendship and support.

At Palgrave Macmillan, we have many to thank. On the editorial side, we are grateful to editor Nicola Jones for championing our manuscript, to associate editor Lani Oshima for her early stewardship of the project, to designer Paileen Currie for the wonderful cover, and to editorial assistants Elaine Fan and Mara Berkoff for answering our many queries and always with such good cheer. Jeff LaSala ably managed the production of the book, while Deepa John at Newgen Knowledge Works oversaw

xiv

the copy-editing and typesetting of the book. Chris Schuck's proofreading of the final draft was invaluable. Finally, we both count our blessings for our children: Brendan McGough, Abby McGough, Rabi Whitaker, Zoey Whitaker, and Dylan Whitaker, and the newest addition to the Whitaker family, Zinnia Whitaker, who turned one-year-old on the day this acknowledgment was written.

PART I

Seeds of Corruption

CHAPTER ONE

A Case Study of Institutional Corruption

Institutional corruption, as a field of inquiry, does not shy away from challenging powerful institutions—from Congress and the professions to the academy itself...and it leads us to interrogate them in important ways.

—Jonathan H. Marks, 2012[1]

In a democratic society, we hope and expect that institutions that serve a public interest will adhere to ethical and legal standards. However, in recent years, we have seen numerous institutions failing to meet that obligation. Greed on Wall Street nearly led to the collapse of our public banking system. Congress is beholden to special interests. We have seen a religious institution, the Catholic Church, systematically fail to protect children from sexual abuse. What these scandals share in common is that they cause social harm and erode the public's faith in its institutions, and thus weaken the democratic core of society.

In this book, we put the spotlight on a medical discipline, organized psychiatry, that has transformed our society in the past 30 years.

Our society thinks of medicine as a noble pursuit, and thus it expects a medical profession to rise above financial influences that might lead it astray. The public expects that medical researchers will be objective in their design of studies and their analysis of the data; that the results will be reported in an accurate and balanced way; and that the medical profession will put the interests of patients first. However, in recent years, there has been a steady flow of reports, both in the mainstream media and in academic journals, detailing the corrupting influence that pharmaceutical money has had on modern medicine. In a 2009 essay, Daniel Wikler, a professor of ethics at Harvard School of Public Health, wrote of how this is undermining societal confidence in the medical profession:

Erosion of medical integrity is not a mere detail, but rather strikes at the heart of what it is to practice medicine. The basis for medicine's claim to be a profession rather than a trade, exchanging a degree of self-governance and autonomy to be trusted experts, is the assurance that this trust will not be misplaced.[2]

Or as Giovanni Fava, editor of *Psychotherapy and Psychosomatics*, wrote in 2007: "The issue of conflicts of interest in medicine has brought clinical medicine to an unprecedented crisis of credibility."[3]

While conflicts of interest may bedevil many medical disciplines, psychiatry is often seen at the epicenter of this "crisis of credibility." In a 2000 editorial titled "Is Academic Medicine for Sale," *New England Journal of Medicine* editor Marcia Angell told of how her journal, when it sought to find an expert to write a review of an article on the treatment of depression, "found very few who did not have financial ties to drug companies that make antidepressants."[4] More recently, in 2008, Senator Charles Grassley sent a letter to the American Psychiatric Association (APA), expressing his concerns about industry influence over the APA and all of organized psychiatry. The social peril is that undue industry influence may be compromising organized psychiatry's public health mission in subtle but far-reaching ways.

Although our society today is focused on the possible corrupting influence of pharmaceutical money on medical disciplines, it is important to recognize that guild interests may also affect their behavior. As Kerianne Quanstrum, a physician at the University of Michigan, noted in a 2010 letter in the *New England Journal of Medicine*:

> Although it is true that individual medical providers care deeply about their patients, the guild of health care professionals—including their specialty societies—has a primary responsibility to promote its members' interests. Now, self-interest is not in itself a bad thing; indeed it is a force for productivity and efficiency in a well-functioning market. But it is a fool's dream to expect the guild of any service industry to harness its self-interest and to act according to beneficence alone—to compete on true value when the opportunity to inflate perceived value is readily available.[5]

Like all medical specialty organizations, the APA is not immune to such ethical challenges. It may be characterized as both a professional organization and a guild, and in this latter function it is expected to further the interests of the profession and its members. Thus, it may feel a strong need to promote a societal belief in the merits of treatments prescribed by psychiatrists, even if science is raising questions about those treatments. As a result, the APA may struggle to provide unbiased information and guidance for the treatment of mental disorders. APA's guild interests may conflict with its public health mission, providing a fertile ground for institutional corruption.

Such worries raise a profound set of questions that, given psychiatry's pervasive influence over our society today, need to be answered. Over the past 35 years, has organized psychiatry met its societal obligation to conduct objective research? Has it disseminated fully accurate information on the efficacy and safety of psychotropic medications? Has it produced truly evidence-based diagnostic and clinical practice guidelines? And if organized psychiatry has not fully met those ethical obligations, what has been the resulting social harm?

Conceptually, we see the APA and academic psychiatry as the twin pillars of organized psychiatry in the United States, and thus jointly the "institution" we are studying. Together, they produce the diagnostic and treatment guidelines that determine psychiatric care in the United States. The psychiatric nosology that the APA produces, the *Diagnostic and Statistical Manual of Mental Disorders* (DSM), is often referred to as the "bible" of psychiatric disorders; the nomenclature, criteria, and standardization of psychiatric disorders codified in the DSM affect a diverse set of areas ranging from insurance claims to jurisprudence. Moreover, through its relationship to the International Classification of Diseases, the system used for classification by many countries around the world, the DSM has a global reach.

For this reason, the influence of American psychiatry extends beyond US borders. Numerous other societies—particularly in Western Europe, Canada, Australia, and New Zealand—have embraced the medical model of care promoted by American psychiatry, and this has led to a marked increase in the diagnosis of mental disorders in those societies too, and in their use of psychiatric medications. Much of the Western world has reason to inquire about the ethics of organized psychiatry.

Studying Institutional Corruption

This book arose out of a year that we, the two authors of this book, spent as fellows at the Edmond J. Safra Center for Ethics at Harvard University. The center has sought to develop an intellectual framework for identifying, investigating, and understanding institutional corruption, with the hope that this process ultimately illuminates solutions for remedying the problem. We are exploring this topic, the ethics of modern psychiatry, through that ethical lens.

First and foremost, institutional corruption is to be distinguished from individual corruption. When we speak about "corrupt behavior," we usually think about how individuals may have acted in immoral or even illegal ways, and have done so for personal gain. That is a story of "bad people" doing "bad things." In those instances, individuals within an institution may have behaved in a corrupt manner, but the institution itself, in its "normative" practices, may not have become corrupt. For instance, we think of Bernie Madoff as a financial manager who acted in a criminal manner, that is, outside the norm, and thus we don't necessarily see, in his behavior, evidence of institutional corruption within the larger financial investment community.

Institutional corruption is different. The problem does not arise from a few corrupt individuals who are hurting an organization, even though the organization's integrity remains intact. Instead, as Lawrence Lessig, director of the Edmond J. Safra Center for Ethics has observed, institutional corruption refers to the systemic and usually legal practices that cause the institution to act in ways that undermine its public mission and effectiveness, and ultimately weaken public trust in the institution. There is a "bending" of the original mission, which results from the normalization of behaviors that compromise truth-seeking. Such practices and behaviors may arise when there are financial incentives at work that encourage the members of the institution to *regularly* behave in ways that undermine the institution's integrity and its capacity to fulfill its public mission. Moreover, the leaders of the organization will likely be unaware that their institution has become "corrupted," and protest against those who have come to see their institutional behavior as being at odds with the organization's intended mission.

Indeed, within this framework of institutional corruption, there is the assumption the individuals within the institution are "good" people. Lessig made this point in his writings on Congress: "These are not bad souls bending the public weal to private ends...We can presume that individuals within the institution are innocent; the economy of influence that they have allowed to evolve is not."[6]

As can be seen, thinking of institutional corruption in this manner focuses attention on the larger social, political, and cultural factors that affect the institution. Institutions exist within a larger socio-political context. Thus, there is a need to identify the "economies of influence" that create the potential for improper dependencies to develop. Whereas individual quid pro quo corruption assumes that there is a "bad apple" problem, the framework of institutional corruption seeks to understand whether there is a "bad barrel" problem. This understanding in turn leads to a framework for identifying proposed solutions, as they must neutralize the financial influences that have corrupted the institution in a systematic way.

A Case Study of Institutional Corruption

We see this book as a case study of institutional corruption. The APA's stated mission is to provide "humane care and effective treatment for all persons with mental disorders," and we examine its past 35-year history to assess whether it has stayed true to that mission. First, we document how two "economies of influence" over the field—pharmaceutical funds and guild interests—arose and became entrenched. Then we look at how these influences have led to a distortion of "scientific truths" and to significant social injury. Next, having laid out this story of institutional corruption, we investigate whether organized psychiatry, on the whole, recognizes that its behavior may have undermined its public mission and led to a loss of the public's trust. This inquiry leads us to delve into the science of "cognitive dissonance," and how difficult it can be for a medical profession, operating under two "economies of influence," to see itself as compromised. Finally, we explore strategies for reform and remedies for institutional corruption in organized psychiatry.

Given this focus, we are not interested in identifying instances where individuals within psychiatry may have behaved in corrupt or fraudulent ways. Instead, we presume that individuals within psychiatry want to serve their patients well, and that they seek to engage in research that will improve treatments for psychiatric disorders.

Ours is an academic inquiry. Yet, we believe it is of critical importance to American society, and, in fact, to many societies around the world. During the past 35 years, psychiatry has transformed American culture. It has changed our view of childhood and what is expected of "normal" children, so much so that more than 5 percent of school-age youth now take a psychotropic drug daily.[7] It has changed our behavior as adults, and in particular, how we seek to cope with emotional distress and difficulties in our lives. It has changed our philosophy of being too, as we have come to see ourselves as less responsible for ourselves, and instead more under the control of brain chemicals that may or may not be in "balance." Our use of psychiatric medications could even be said to range from womb to grave: An increasing number of infants born today are exposed to an antidepressant in utero, and psychiatric drugs are regularly given to the elderly in nursing homes to individuals without psychiatric disorders. As such, our society has a compelling interest in the investigation that is at the heart of this book.

In addition, we hope that our book will serve as a "case study" for investigating institutional corruption and for developing solutions that could neutralize the "improper dependencies" that fostered the corruption. In that way, we hope that it can contribute to the discussion of how democratic societies can encourage its public institutions to behave in ways that serve the public good.

CHAPTER TWO

Psychiatry Adopts a Disease Model

The struggle over the drafting and publication of the DSM-III appeared to be a clinical debate, but underlying it all was a vehement political struggle for professional status and direction.

—Rick Mayes, 2005[1]

The modern era of psychiatry, in terms of its classification of mental disorders, dates to 1980, when the American Psychiatric Association (APA) published the third edition of its *Diagnostic and Statistical Manual* (*DSM-III*). This marked the moment that the APA moved away from psychoanalytic explanations for mental disorders and adopted what it considered to be a more scientific way to think about psychiatric difficulties. The disorders were to be diagnosed based on characteristic "symptoms," a model that other medical specialties, when faced with illnesses of unknown causes, had long used. The public was encouraged to think of mental disorders as "diseases," and very soon, with this concept in mind and the arrival of new psychiatric medications on the market, the use of these drugs soared. In the United States, spending on psychiatric drugs increased from around $800 million in 1985 to more than $40 billion in 2011, evidence of how the diagnosis of mental disorders and the prescribing of psychiatric medications have dramatically expanded since the publication of *DSM-III*.[2]

Since our focus is on the behavior of the APA and organized psychiatry in this modern era, we need to begin our inquiry with a review of the various forces that led to the creation of *DSM-III*. It is easy to see both a scientific impulse and guild impulse at work, and in that second impulse, the seed for possible corruption of the scientific enterprise.

The Historical Path to DSM-I

The history of psychiatric nosology in the United States is a complicated one, partly because mental and emotional difficulties appear in such disparate forms, and partly because of the long-running philosophical argument over whether they are biological or psychological in nature. The architects of *DSM-III* spoke of wanting to "remedicalize" the field, which reflects the fact that the diagnostic approach they adopted—classifying disorders based on symptoms—was not a novel idea, but rather one that had been employed in psychiatric texts before.

In 1812, Benjamin Rush, who is remembered today as the father of American psychiatry, authored what might be considered the first US text on mental disorders. In *Medical Inquiries and Observations upon the Diseases of the Mind,* Rush sought to classify psychiatric illnesses, and he theorized about their possible causes and proper treatment. Rush believed that madness was "seated in the blood vessels," and that, broadly speaking, mania, delusions, and other agitated states were the result of too much blood flowing to the head, while states of inactivity and torpor were the result of too little blood to the brain. His recommended treatments were designed to repair this circulatory imbalance, either by increasing or decreasing blood flow to the head.

Shortly after Rush's book appeared, Quakers in the United States began building small asylums in pastoral settings for treatment of the "mad." Whereas Rush, who had studied at the University of Edinburgh, conceived of mental disorders as medical illnesses, the Quakers, steeped in a religious approach, professed not to know the causes of madness, but they were certain that the mad should be treated as "brethren." Their pastoral retreats were designed to be places that could "assist nature" in helping the mad heal.

Even so, a medical approach to treating madness was not altogether absent from these asylums. Physicians were hired to serve as superintendents of the small private asylums, and they professed an interest in trying to identify different forms of insanity. In 1844, 13 asylum physicians formed the Association of Medical Superintendents of American Institutions for the Insane, which would later evolve into the American Psychiatric Association, and one of the group's first discussions focused on the classification of mental disorders. But when they looked around their wards, they did not see patients who could easily be grouped together into diagnostic types. "Insanity," said the association's first president, Samuel Woodward, is "easily recognized...[but] not always easily classified."[3]

In the second half of the nineteenth century, the patient pool in asylums grew even more diverse. In the 1840s and 1850s, the reformer Dorothea Dix, impressed with the kind care offered by the small Quaker asylums, urged states to build such pastoral retreats for the general public. States responded to her entreaties in an admirable way, providing funding for this endeavor, and the number of mental hospitals in the United States, private and public, increased from 18 in 1840 to 139 in 1880.[4] However, as these new asylums were built, communities began dumping people with all kinds of illnesses into them. Syphilitics, alcoholics, the mentally disabled, and the senile elderly joined the newly insane in these hospitals.

In 1885, a small group of alienists, as psychiatrists were then called, identified eight categories of insanity: mania, melancholia, monomania, dementia, general paralysis of the insane, epilepsy, toxic insanity, and congenital mental deficiency.[5] While several of these forms of madness were clearly due to physical illnesses, diagnosis still depended on identifying differences in symptoms, noted Pliny Earle, one of the founders of AMSAII. "In the present state of our knowledge, no classification of insanity can be erected upon a pathological basis, for the simple reason that, with but slight exceptions, the pathology of the disease is unknown. Hence, for the most apparent, the most clearly defined, and the best understood foundation for a nosological scheme for insanity, we are forced to fall back upon the symptoms of the disease—*the apparent mental condition*, as judged from the outward manifestations."[6]

In Europe, German psychiatrist Emil Kraepelin soon provided psychiatry with a more developed scheme for classifying functional psychotic disorders (i.e., psychosis without an apparent somatic cause), one that grew out of his study of the long-term course of psychotic patients at a mental hospital at the University of Dorpat in Estonia (the University of Tartu today). Those psychotic patients who presented with affect—mania, depression, etc.—were likely to get better, and their long-term course was fairly benign. These patients, he said, suffered from manic-depressive illness. Those psychotic patients who presented with a lack of affect, and who often exhibited a striking physical deterioration (and had difficulty making willed movements, much like Parkinson's patients), would likely become chronically ill and experience early dementia. These patients, he said, suffered from *dementia praecox*.

Kraepelin's work marked a clear step forward for the field. His was a classification system that, at least for psychotic patients, linked presenting symptoms with clinical course. For that reason, Kraepelin provided psychiatry with its first classification system that was both reliable, in that these two types of patients could be distinguished with some consistency, and valid, in that the differential diagnosis led to a different prognosis.

However, only a fraction of the patients committed to US mental hospitals in the early 1900s fell neatly into one of these two categories. The hospital wards were also crowded with demented patients, epileptics, imbeciles, and patients ill with syphilis, Huntington's chorea, alcoholic psychosis, and other such ailments. Diagnostic uncertainty reigned. In 1908, the US Census Bureau, which for 70 years had regularly sought to tally up the number of insane people in the country, asked the American Medico-Psychological Association to help with the gathering of data on the nation's mentally ill (AMSAII had evolved into the American-Medico Psychological Association by this time). The association, as it contemplated this request, admitted that it lacked a proper diagnostic manual. "The present condition with respect to the classification of mental diseases is chaotic," it confessed. "This condition of affairs discredits the science of psychiatry and reflects unfavorably upon our association."[7]

In 1918, the American Medico-Psychological Association, in collaboration with the National Committee for Mental Hygiene, issued the field's first standardized psychiatric nosology, a *Statistical Manual for the Use of Institutions for the Insane.* The manual divided mental disorders into 22 principal groups. Twenty were for disorders that were presumed to have a biological cause (syphilis, alcoholism, Huntington's chorea, cerebral arteriosclerosis, etc.) and the remaining two groups were for disorders without a presumed biological cause: one for psychotic patients and the other for disturbed patients without psychosis.

For the next two decades, this manual, with its categorization of institutionalized patients, provided psychiatry with an adequate nosology. Psychiatry was a discipline firmly rooted in the asylum, such that by the beginning of World War II, more than two-thirds of the APA's 2,300 members still worked in mental hospitals.[8] However, during the war, psychiatrists and other physicians assigned to neuropsychiatric units treated soldiers traumatized by battle. In that setting, the physicians discovered that psychotherapy plus rest often worked miracles; 60 percent of soldiers identified as "neuropsychiatric casualties" returned to duty within five days.[9] The war physicians saw that people could descend into a highly disturbed state and quickly become well again, which was an experience that psychiatrists working in asylums did not frequently have.

"Our experiences with therapy in war neuroses have left us with an optimistic attitude," Roy Grinker wrote. "The lessons we have learned in the combat zone can well be applied in rehabilitation at home."[10]

This wartime experience led many psychiatrists to reconsider the nature of mental disorders. Clearly, environmental stresses and other psychological challenges could trigger symptoms of mental illness. This was a conception that fit with Freudian conceptions of mental disorders, which were being popularized after the war. In 1946, William Menninger, a psychiatrist who had risen to the rank of brigadier general during the war, and a group of younger psychiatrists formed the Group for the Advancement of Psychiatry, which sought to remake the field. They saw a need for American psychiatry to help people in the community who were not so seriously ill that they needed to be hospitalized, but were struggling with anxiety, or other "neurotic" conditions.

With many in the profession now adopting Freudian beliefs and focused on treating a broader group of patients, the field clearly needed a new diagnostic text. "Current nosologies and diagnostic nomenclature," Menninger argued, "are not only useless but restrictive and obstructive."[11] In 1950, the APA formed a committee to create a new nosology, and two years later, the APA published the first edition of its *Diagnostic and Statistical Manual (DSM-I).*

A spiral bound, 130-page book, *DSM-I* divided patients into two broad groups. The first was composed of patients whose mental disturbances arose from an evident biological cause: infection, poisons, alcohol intoxication, circulatory illnesses, metabolic problems, brain cancers, multiple sclerosis, and other hereditary diseases were common ones. These patients were familiar to asylum doctors, and such diagnostic categories could have been found in the *Statistical Manual for the Use of Institutions for the Insane*. What was new in *DSM-I* was the second broad diagnostic grouping. It identified disorders resulting from the individual's inability to adjust well to his or her environment. This second diagnostic group, which clearly reflected Freudian ideas, was then subdivided into psychotic and psycho-neurotic disorders. The psychotic conditions included manic-depressive, paranoid, and schizophrenic "reactions." The psychoneurotic disorders were diagnoses for people in the community who struggled with anxiety, obsessive-compulsive behavior, depression, emotional instability, and other such problems.

Ever since Rush had published his *Medical Inquiries and Observations upon the Diseases of the Mind,* the field had mostly relied on a "medical model" for classifying mental disorders. Diagnosis was based on presenting symptoms. The one exception to this practice had been during the first half of the nineteenth century, when the moral therapy pioneered by the Quakers was popularized. But with the publication of *DSM-I*, the APA had moved away from a medical orientation, particularly when it came to diagnosing patients living in the community. Anxiety, depression, and even psychosis were not to be seen as symptoms of a disease, but rather emotional distress that arose from internal psychological conflicts and the particulars of a person's life story. There was no longer a clear line that divided the mentally well from the mentally ill, and, if anything, it seemed that unresolved psychological conflicts probably plagued most people, at least to some degree.

Psychiatry now had a diagnostic text that enabled it to move out of the asylum and provide care for a larger segment of the American population. This transformation happened quickly; by 1956, only 17 percent of the APA's 10,000 members were employed in mental hospitals.[12] The couch, employed as part of a Freudian talking cure, was the new symbol of psychiatry's workplace, as opposed to the asylum. "It does not seem possible that Kraepelin so recently dominated psychiatry," Harvard Medical School psychiatrist Stanley Cobb marveled in 1961, in an *Atlantic Monthly* article titled "Psychiatry in American Life."[13]

Psychiatry's Reliability Problem

Many of the psychoanalysts who came to dominate American psychiatry in the 1950s did not see diagnosis as particularly important. The etiology of all nonorganic mental disorders was presumed to be the same—a psychological failure to adapt to one's environment—and psychotherapy was a treatment that was supposed to help remedy that failure. As such, there was no need to put "so much emphasis on different kinds and clinical pictures of illness," Menninger wrote.[14] Still, psychiatry was a branch of medicine, which understood good diagnostics as essential to good care, and not everyone in the APA agreed with Menninger that diagnostic categories were not particularly important. When *DSM-I* was introduced, George Raines, chair of the APA committee that produced the manual, declared that "accurate diagnosis is the keystone of appropriate treatment and competent prognosis."[15] At the very least, Raines and others wanted the manual to be a reliable method for sorting psychiatric patients into different categories, which could then facilitate research into those diagnoses.

During the next decade, it became evident that *DSM-I* did not provide psychiatry with a taxonomy that fulfilled that goal. In 1962, Aaron Beck reviewed nine studies assessing reliability of functional disorders (those without an apparent organic cause), and found that diagnostic agreement among psychiatrists ranged from 32 to 42 percent, which was little better than chance.[16] Five years later, Robert Spitzer, who would soon be named to head the task force that would develop *DSM-III*, introduced into psychiatry a new mathematical method for assessing diagnostic reliability, called kappa scores, and he subsequently determined, in a review of the scientific literature, that "the reliability of psychiatric diagnosis as it has been practiced since at least the late 1950s is not good."[17]

Psychiatry's reliability problem burst into public view in 1973, when Stanford University psychologist David Rosenhan reported on a novel experiment he had conducted. He and seven of his students had shown up at 12 different mental hospitals (some went to more than one hospital), complaining that they heard voices, vague in nature, which said such things as "thud," "empty," or "hollow." Those were the only symptoms they gave, and yet in every instance they were admitted to the hospital, and in every case but one, they were diagnosed as ill with schizophrenia. Once admitted, the pseudopatients stopped complaining of any symptoms and behaved normally, but in spite of this, none of the hospital staff ever spotted them as imposters. The only ones in the hospital who didn't fall for their ruse were the "real" patients in the hospital. "The facts of the matter are that we have known for a long time that diagnoses are often not useful or reliable, but we have nevertheless continued to use them," Rosenhan wrote in *Science*. "We now know that we cannot distinguish insanity from sanity."[18]

Although Spitzer and the APA's leadership bristled at this conclusion, the field nevertheless acknowledged that, with the use of *DSM-I* and then *DSM-II* (published in 1968), it had a reliability problem. Now it had erupted into the public domain, and this complaint—that psychiatry lacked a way to reliably distinguish between normal behavior and illness—began to be regularly voiced by insurance companies and in government circles, too. As US Senator Jacob Javits declared in 1977: "Unfortunately, I share a congressional consensus that our existing mental health care delivery system does not provide clear lines of clinical accountability."[19]

In other areas of medicine, it was understood that in order for a diagnostic manual to be useful, it should enable physicians to regularly come to the same diagnosis, and that different diagnoses should be valid, meaning that the underlying pathology of an illness had become known, or that at least research had identified a characteristic course for the illness, which distinguished one diagnosis from another. *DSM-I* and *DSM-II* hadn't provided psychiatry with a reliable diagnostic manual, and as Spitzer and others knew, the disorders in the manual couldn't be said to be validated either. Although Freudian theories posited an etiology for mental disorders—that they arose from unconscious conflicts in the mind and a failure to adapt to stressful environments—these theories couldn't be empirically tested. From a medical perspective, psychiatry's existing taxonomy was inadequate and needed to be rethought.

The Scientific Impulse for Reform

Although psychiatrists schooled in psychoanalysis or psychodynamic therapy led most psychiatry departments at US medical schools in the 1950s, there were still a small number of psychiatrists who remained loyal to Kraepelinian ideas that serious mental disorders were discrete illnesses. In this classification scheme, schizophrenia was seen as different from manic-depressive illness as tuberculosis was from malaria. The premier academic home for such contrarians was Washington University in St. Louis, where the department was led by Eli Robins, Samuel Guze, and George Winokur. The Washington group derided Freudian concepts as "unscientific" and lacking in empirical support, and argued that *DSM-I*, because of its reliability problem, had hindered research into the etiology of psychiatric disorders and how they might be better treated.

In 1967, the Washington University group began meeting regularly to discuss developing new diagnostic criteria, for use in research, and they took Kraepelin as their guide. Kraepelin had linked different presenting symptoms to different diagnoses, which had differing prognoses, and he had done so by following patients for a longer period of time. The Washington University group decided that for a diagnostic category to have some validity, it should provide a detailed clinical description of presenting symptoms, with explicit criteria for making a diagnosis, and it should be informed by research. The necessary science included laboratory studies of patients with the presenting symptoms; studies of their long-term outcomes; and family studies, to determine whether there might be a genetic component. The Washington University group sought to develop diagnoses supported by empirical data, rather than solely by clinical wisdom.[20]

With these validity criteria in mind, a psychiatric resident at Washington University, John Feighner, scoured the medical literature for evidence that could support psychiatric diagnoses. He found such evidence, even to a limited degree, for only 16 psychiatric illnesses. When Feighner published his paper, he noted that this was but a first step for the field. He and his Washington University collaborators had not conducted any long-term studies of their own. They had not performed studies to assess whether the explicit criteria for making a diagnosis—for instance, requiring that five of eight symptoms said to be characteristic of depression be present before making the diagnosis—was meaningful. They had developed the diagnostic criteria in order to group patients into homogeneous samples for research, and not as a diagnostic scheme for clinicians. "All [these] diagnostic criteria are tentative in the sense that they [will] change and become more precise with new data," Feighner wrote.[21]

There was a clear scientific reasoning in Feighner's work. In order to study patients who presented with differing symptoms, the field needed diagnostic categories that would provide adequate "reliability" and might also have at least a limited "validity," arising from studies that included data on long-term outcomes (which was the data that Kraepelin had relied upon to divide psychotic patients into two broad groups). But given the limitations of existing data on psychiatric disorders, the Washington University group understood that considerable research would be required before the field had a diagnostic manual that clinicians could use, confident that it was both reliable and valid.

Psychiatry's Legitimacy Crisis

Even as Rosenhan's study illuminated the fact that psychiatry had a reliability problem, the field was also having to confront two much larger problems: societal questions about its legitimacy, and competition in the marketplace for patients. Together, these twin problems created a crisis for the field in the early 1970s, and, ultimately, provided the APA with powerful guild reasons for redoing its diagnostic manual.

For a long time, psychiatry's public image had been a muddled one. In the late 1800s, the psychiatrists were "superintendents" of mental hospitals, and thus they were often seen more as caretakers of those institutions, as opposed to physicians who treated diseases. Then, in the first half of the twentieth century, they practiced in an environment that was mostly forgotten about by the American public. In the late 1930s and 1940s, there were a flurry of reports in the media about new wonder treatments for the mad—insulin coma therapy, convulsive therapies, and finally frontal lobotomy—but that good press came to an abrupt halt at the finish of World War II, when the American public was shocked by newspaper and magazine exposes of the horrendous conditions in many of the nation's mental hospitals. In his book *The Shame of the States*, Albert Deutsch even compared what he had seen in such institutions to what he had witnessed when the Nazi concentration camps were liberated. More than ever, psychiatry was seen as a medical backwater, a discipline that needed to be remade and provided with a new scientific foundation. In 1946, Congress created the National Institute of Mental Health (NIMH) to direct this task.

That postwar era of shame soon turned into a decade of growth and success. During the 1950s, Freudian theories of the mind captured the American imagination, and psychiatry also regularly announced the discovery of new drugs that were remaking the field. In 1954, chlorpromazine came to market, and *Time* magazine described this major tranquilizer—which was soon to be renamed an antipsychotic—as a "wonder drug."[22] Next the magazines were telling of a miraculous new pill, iproniazid, for rousing depressed patients, and then Miltown, a drug for anxiety, took the nation by storm. This drug was dubbed a "happiness" pill, which, *Time* reported, was for "walk-in neurotics rather than locked-in psychotics," and demand was so great that pharmacies, when they were able to have it in stock, would put up announcements that shouted, "Yes, we have Miltown!"[23]

Psychiatry was suddenly a popular discipline. From 1946 to 1956, the number of slots at medical schools for psychiatry residents quadrupled, from 758 to 2,983, and a long-spurned specialty began to attract some of the top medical students.[24] Moreover, during the 1950s, psychiatrists who treated patients in the community saw their incomes notably increase. Whereas an asylum psychiatrist, by the end of the decade, was earning on average only $9,000 per year, the Freudian analyst in private practice was earning more than twice that, $22,000 per year. The top-earning psychiatrists were those with a biological orientation who focused on prescribing the new wonder drugs to walk-in patients. They were making $25,000 a year.[25]

Yet, as psychiatry moved out of the asylum and expanded its influence over American society in the 1950s, new criticisms arose. There was a faction within psychiatry who began arguing that the field needed to address the many social problems—poverty, etc.—that could stir mental distress, but, critics asked, if social factors were the source of many psychiatric ailments, then why was treating mental illness a task for psychiatry? Wasn't it the responsibility of society to fix such problems? Psychological explanations for severe mental illnesses, such as schizophrenia, also raised new questions about the legitimacy of mental hospitals. Were they places for the "ill," or simply warehouses for rebels, misfits and malcontents whose behaviors, formed by reactions to a difficult environment, could be offensive to others? Did the hospitals function as institutions for social control?

The opening salvo against psychiatry was fired in 1961 by Thomas Szasz. In his book *The Myth of Mental Illness*, he dismissed the idea of psychiatric disorders as "scientifically worthless and socially harmful."[26] Psychiatrists, he said, functioned more as police and ministers than as medical doctors. His book was favorably reviewed by mainstream publications, such as the *Atlantic* and *Science*, evidence that his critique tapped into a larger societal concern. *Science* described his book as "enormously courageous and highly informative...bold and often brilliant."[27]

Critiques of psychiatry, mostly arising from the halls of academia, gained a larger readership, and together these writers stirred an "antipsychiatry" movement. Sociologist Erving Goffman, in *Asylums*, argued that mental hospitals infantilized people and turned them into chronic patients. David Cooper, R. D. Laing, and others wrote similar criticisms. In the world of fiction, Ken Kesey and his book *The Cuckoo's Nest* introduced America to Randle McMurphy, a rebel and rogue who had been hospitalized because of his anti-authoritarian ways. In the early 1970s, a civil rights lawyer, Bruce Ennis, penned *Prisoners of Psychiatry,* a title that told all about the societal function he believed psychiatry served.

At this time, ex-patients also began organizing, dubbing them-selves "survivors" of psychiatry. The Insane Liberation Front formed in Portland; the Mental Patients' Liberation Project in New York City; and the Network Against Psychiatric Assault in San Francisco. They held demonstrations, organized human rights conferences, and spoke of their struggle as a fight for civil rights. They described antipsychotics as tools of suppression, rather than as healing agents, and even took their fight to state courts, where they argued that forced drug treatment was a form of medical assault, and a violation of their constitutional rights to due pro-cess and freedom of speech. Gay rights groups added their voices to the mix, demanding that psychiatry remove homosexuality from its diagnos-tic manual. Yet, when the APA voted to amend this diagnosis in 1974, creating in its place a disorder called ego dystonic homosexuality, which was to be diagnosed only if being homosexual caused one distress, the critics of psychiatry had another round of ammunition: were psychiatry's diagnoses so uncertain that they could be changed in response to polit-ical pressure?

Even as psychiatry's public image was being battered, it was fighting in the marketplace with psychologists, social workers, and counselors over control of psychotherapy. Psychiatry had been battling psychologists since the end of World War II, when psychologists, who previously had focused on developing mental tests and studying the human personality, began providing clinical services. Their move into the psychotherapy market accelerated after Carl Rogers published *Client-Centered Therapy* in 1951. In response, the American Psychiatric Association, with support from the American Medical Association, sought to block psychologists by invok-ing state medical licensing laws, which prohibited nonphysicians from practicing medicine. Clinical psychologists could provide psychotherapy services only under the supervision of a medical doctor, the APA argued. "Psychotherapy is a medical treatment and does not form the basis of a separate profession," the APA stated in 1958.[28]

Throughout the 1950s and 1960s, psychologists waged a state-by-state struggle to get out from under psychiatry's yoke, seeking state licensure to provide psychotherapy without being supervised. One by one, the states agreed. Psychotherapy was not a "medical procedure" that could be pro-vided only under the supervision of a physician, they decided. By 1962,

18 states had granted psychologists the right to practice psychotherapy without such supervision, and by 1977, every state had.[29]

Even worse for psychiatry, competing therapists were popping up in every corner of American society. By 1980, there were more clinical psychologists in the United States than there were psychiatrists (50,000 to 28,000), and there was a much greater number of social workers, family therapists, marriage counselors, and other "healers" offering help to American's walking wounded.[30] Psychiatrists providing talk therapy were in competition for patients with a multitude of professionals and even nonprofessionals, and what was particularly galling for the field was that it lacked evidence that its preferred brand of psychotherapy, psychoanalysis, was any more effective than the many other types of psychotherapy that were being offered.

This finding had first surfaced in 1952, when Hans Eysenck, a British behaviorist, reported that two-thirds of neurotic patients improved, regardless of how they were treated, or whether they were treated at all. Other studies found that while psychotherapy may be better than no treatment at all, no specific treatment was better than another, leading one scholar to later quip that such research had produced a "Dodo bird's verdict," in which "everybody won and all must have prizes."[31]

Given this fact, medical insurance companies began questioning why they should provide coverage for talk therapy. From 1965 to 1980, the percentage of Americans with private medical insurance rose from 38 to 68 percent and, as the result of the enactment of Medicare and Medicaid in 1965, federal and state expenditures for mental services also rose dramatically during this period.[32] Both the private insurers and the government were beginning to perceive psychiatry, APA medical director Melvin Sabshin later recalled, as a financial "bottomless pit," uncertain whether their rising expenditures were producing a benefit.[33] The problem, complained a Blue Cross executive in 1975, was that "compared to other types of [medical] services, there is less clarity and uniformity of terminology concerning mental diagnoses, treatment modalities and types of facilities providing care."[34]

In 1977, Harvard Medical School psychiatrist Thomas Hackett summed up psychiatry's understandable anxiety over its place in the talk therapy marketplace: "Apart from their training in medicine, psychiatrists have nothing unique to offer that cannot be provided by psychologists, the clergy, or lay psychotherapists. Our bread and butter—the practice of psychotherapy—has fragmented into multiple schools, all with uncertain boundaries."[35]

Such were the many challenges that confronted psychiatry in the late 1960s and 1970s. An antipsychiatry movement challenged its legitimacy. Ex-patients told of having "survived" psychiatric hospitals and drug treatments. Psychiatrists providing talk therapy were now in competition with an ever-growing multitude of therapists tending to the psychic wounds of Americans. Insurance companies were balking at paying for mental health services. Its own reliability studies revealed that its diagnostic practices were problematic. Efficacy studies hadn't shown that its brand of psychotherapy was any better than that offered by its competitors. Faced with so many problems, APA leaders spoke of how psychiatry was under "siege," and a few prominent psychiatrists even worried that psychiatry, as a field, could be headed for "extinction."[36]

The Guild Impulse for Remaking the DSM

Ever since *DSM-II* had been published in 1968, the APA had a bureaucratic reason to at least slightly amend it. The United States was a member of the World Health Organization, which required that the APA's classification of mental disorders be compatible with the taxonomy set forth in the WHO's *International Classification of Diseases (ICD)*. *DSM-II*, because of its Freudian description of psychotic disorders with unknown somatic causes as "reactions," was incongruent with the *ICD*, which assumed that all psychotic disorders had a biological cause, even if the cause was unknown. The reliability problem with *DSM-I* and *DSM-II* provided a scientific rationale for redoing it as well. But, as historian Hannah Decker noted in her book *The Making of DSM III*, it was Rosenhan's report in *Science*, detailing how he and his quite-normal students had been diagnosed as schizophrenic, that prompted the APA to push this initiative to the front burner.

Shortly after Rosenhan's article was published, the trustees of the APA convened a meeting in Atlanta. For three days, they discussed the "rampant criticisms" of psychiatry, lamented the fact that the public did not have a "strong conception of psychiatry as a medical specialty," and failed "to recognize a psychiatrist's special competence in mental health care."[37] At the end of their meeting, the trustees decided that *DSM-II* should be fundamentally revised, and urged that this task be completed in two years. The trustees also recommended the formation of a task force that would "define mental illness and what is a psychiatrist," which could then be used as a preamble to *DSM-III*.[38]

At this moment of *DSM-III*'s conception, the APA trustees saw that creating a new diagnostic manual could serve a guild interest. Freudian ideas had shaped *DSM-I* and *DSM-II*, but that diagnostic approach had ultimately produced a crisis for psychiatry. Remaking psychiatric diagnoses could be part of a larger effort by psychiatry to put forth a new image, which, metaphorically speaking, would emphasize that psychiatrists were doctors, and that they treated real "diseases."

Over the next six years, as the *DSM-III* task force labored to produce the new manual, Spitzer, Samuel Guze, APA medical director Melvin Sabshin, and others sounded this theme again and again. *DSM-III*, Spitzer said, would serve as a "defense of the medical model as applied to psychiatric problems."[39] Sabshin told the APA members that a "vigorous effort to remedicalize psychiatry should be strongly supported."[40] Harvard psychiatrist Seymour Kety argued that the "medical model is as appropriate for the major psychoses as it is for diabetes."[41] "The basic premise," noted Arnold Ludwig, a psychiatrist at the University of Kentucky, is "that the primary identity of the psychiatrist is as a physician."[42] Guze, in an article titled "Nature of Psychiatric Illness: Why Psychiatry is a Branch of Medicine," wrote that, with the new model, the focus would be on "the symptoms and signs of illness . . . the medical model is clearly related to the concept of disease." Psychiatrists, he wrote, would now focus on reducing those symptoms, which was a task that psychiatrists could do best. "Medical training is necessary for the optimal application of the most effective treatments available today for psychiatric patients: psychoactive drugs and ECT."[43]

This last sentiment told of a financial incentive lurking in the background as the APA remade its diagnostic manual. Adopting a disease model would lead to a focus on treatments that allayed symptoms, and it would only be psychiatrists, thanks to their prescribing powers, that could provide patients with access to psychiatric drugs. Psychiatry might cede talk therapy to its competitors, but it would have this corner of the therapeutic marketplace to itself. A 1975 survey found that there were very few psychiatrists who didn't prescribe drugs, and that "psychiatrists almost routinely prescribed drugs for patients who were treated by other mental health professionals not licensed to administer drugs."[44] Psychiatry was following a financial path to this role in the therapy marketplace, and adopting a medical model that focused on the "symptoms of a disease" would obviously enhance the value of psychiatrists' prescribing powers.

The Making of *DSM-III*

By naming Robert Spitzer to head the task force, the APA trustees could expect that he would take the profession in a new direction. Although Spitzer, after graduating from NYU School of Medicine, had subsequently trained as a psychoanalyst, he had never really embraced Freudian ideas, and, in his position at the New York Psychiatric Institute in New York City, he had embraced the neo-Kraepelin ideas of the Washington University group. He had written several papers criticizing the reliability of *DSM-I* and *DSM-II*, and he had followed up on Feighner's work to create his own Research Diagnostic Criteria, which had expanded Feighner's list of 16 categories of psychiatric illnesses to 25.[45]

As Spitzer set up his task force, he picked others who shared his belief that Freudian ideas needed to be abandoned. More than half of the task force members had a current or past affiliation with Washington University in St. Louis. At least at first, the task force looked to Feighner's criteria as a standard for creating a diagnostic category. "There will be fewer assignments to diagnostic categories on the basis of probable correctness, and more diagnoses which force the clinician to admit what he does not know," Spitzer said, adding that "the sense of the committee is that mental disorder should be defined narrowly rather than broadly, that a definition which permits false negatives is preferable to one that encourages false positives."[46] Moreover, Spitzer acknowledged that the clinical utility of the new manual was certain to be quite limited, given that the diagnostic categories could best be described as "hypotheses" that would need to be further researched before they could be considered validated. The *DSM-III* criteria, he said in 1975, "would be 'suggested' only, and any clinician would be free to use them or ignore them as he saw fit."[47]

There was a caution and humility in such comments, which arose from an understanding that the field suffered from a lack of high-quality research on psychiatric disorders. The Washington University group, as it developed the Feighner criteria, had noted that long-term studies were needed to create discrete diagnostic categories, and as Guze later recalled, he proposed to the task force that "until there had been at least two long-term follow-up studies from different institutions with similar results, we shouldn't give the entity a status in *DSM III*...that would put us on a stronger scientific basis and it would constantly remind psychiatrists of our ignorance and what kinds of questions needed to be studied."

This was a key moment for the task force. Guze was proposing that the *DSM-III* diagnoses be informed by research, and if such data weren't available, that the group avoid creating a diagnosis. However, his proposal was rejected. "I couldn't get that group to vote in favor of my suggestions," he recalled. "The response that I was given was that they said we have enough trouble getting the legitimacy of psychiatric problems accepted by our colleagues, insurance companies, and other agencies. If we do what you are proposing, which makes sense to us scientifically, we think that not only will we weaken what we are trying to do but we will have given the insurance companies an excuse not to pay us."[48]

Guild interests, it seemed, would have to trump scientific concerns. The task force may have wanted *DSM-III* categories to be based on empirical data, but the science to provide such data hadn't yet been done. "There was very little systematic research" to draw on, said task force member Theodore Millon. "And much of the research that existed was really a hodgepodge, scattered, inconsistent, and ambiguous."[49]

The committee's initial caution soon gave way to a policy of "syndromal inclusiveness," with the thought being that *DSM-III* would provide a diagnosis for all of the patients that psychiatrists now saw. Indeed, Spitzer understood that psychiatry, with its new manual, would be staking a claim to a potential market for its services. "It defines what is the reality," he later said. "It's the thing that says, 'this is our professional responsibility, this is what we deal with.'"[50]

Once this guild interest became paramount, the basis for creating a new diagnostic category became "expert opinion." Spitzer and others would meet, and they would discuss possible criteria for making a diagnosis.

The final criteria, said Allen Frances, who worked on *DSM-III*, "would usually be some combination of the accepted wisdom of the group, as interpreted by Bob [Spitzer], with a little added weight to the people he respected most, and a little bit to whoever got there last."[51]

Millon, in his later review of this process, said that it wasn't just that the APA task force had little good science to draw upon, but that the task force didn't even incorporate the limited empirical research that had been done. As such, it "failed to construct an instrument that reflected previous research."[52] Toward the end of the process, many with a Freudian perspective began to voice objections to the new manual, and that battle further affected the drawing up of diagnostic criteria. "The entire process," Spitzer later confessed, "seemed more appropriate to the encounter of political rivals than to the orderly pursuit of scientific knowledge."[53]

The task force completed a preliminary draft of *DSM-III* by 1976, and beginning in 1977, the APA, with a grant from the NIMH, began field trials of the new manual. Four hundred sixty-seven clinicians assessed the symptoms of 12,667 patients to make a diagnosis based on the new categories. However, the initial phase of the field trials was not done to *assess* reliability, but rather to *improve* it. Spitzer relied on the results and the feedback from the participating psychiatrists to revise the diagnostic criteria so that the categories' reliability would be better, and thus subsequent field trials would produce a better kappa score. The major purpose of this first phase of the field trials, Spitzer told the NIMH, was "to identify and solve potential problems with the *DSM III* draft."[54]

Spitzer and his collaborators then conducted a phase II trial of the redrawn categories, but this study was not particularly scientific in its methodology. People who had been on the task force participated in the trial; the sites that participated were not a representative sample of sites in the country; and there was no comparison made to another classification system, such as *DSM-II*. Although Spitzer published five articles on the field trials, he reported raw data in only one, and there is an inconsistency in the reports on the number of patients that were studied. As Decker concluded in her study of *DSM-III*, given the conflicting data, "it is no simple matter to write about the reliability portion of the NIMH trial."[55] Perhaps reliability would be better with *DSM-III* than with *DSM-II*, and perhaps not; when the methodology of the field trials was carefully assessed, and the published data carefully parsed, it became clear that there was a lack of convincing evidence on the matter.

A Guild Triumph

While there had been a clear scientific impulse behind the APA's remaking of its diagnostic manual, that impulse did not translate into a rigorous scientific process for creating diagnostic categories. On the other hand, the making of this new 494-page manual, which was published in 1980 and listed 265 disorders, did serve the APA's guild interests in a brilliant way.

By adopting a disease model and asserting that psychiatric disorders were discrete illnesses, the APA had addressed both antipsychiatry critiques and its image problem. Metaphorically speaking, psychiatry had donned a white coat. It was presenting itself to the public as a *medical specialty,* which served as a reply to Thomas Szasz and others who argued that psychiatrists functioned "more as police and ministers" than as doctors. This was also an image that resonated with the public. "The medical model," wrote Tufts Medical School psychiatrist David Adler after *DSM-III* was published, "is most strongly linked in the popular mind to scientific truth."[56]

Next, the new manual enabled psychiatry to lay medical claim to the very patients that had enabled the field to leave the asylum. Freudian conceptions of psychiatric distress, including its concept of "neurosis," provided psychiatry with a rationale for treating people with everyday problems, which, in fact, often did arise to such stresses as failing marriages, family conflicts, job difficulties, and other such difficulties in life. However, in the 1970s, psychiatry was losing the battle to treat these patients—the "walking wounded"—with psychotherapy, but *DSM-III,* with its listing of 265 separate disorders, turned the walking wounded into patients with *illnesses,* whose symptoms needed to be treated. That was a model that would support the regular use of psychiatric drugs, which only physicians could prescribe.

At the same time, *DSM-III* solved the profession's difficulties with the insurance companies. As it remade the DSM, Spitzer and others met with the medical directors of major insurance companies, and the medical directors informed Spitzer and the APA that insurance "was meant to pay for the sick, not the discontented who are seeking an improved lifestyle. We need your help in differentiating between those who have mental disorders and those who simply have a problem."[57] If the *DSM-III* task force had followed Guze's lead, then the manual would have drawn an illness boundary that incorporated a much smaller number of people, and the insurance companies would have reimbursed only for the treatment of that smaller group of patients. By insisting that all disorders in the *DSM* were illnesses, the APA could now expect that insurance companies would pay for the treatment of nearly everyone who came to a psychiatrist's office, regardless of the person's problem. The APA, with its new diagnostic manual, had helped the insurance companies differentiate "between those who have mental disorders and those who simply have a problem" by asserting, in essence, that those who had a problem—e.g. the "discontented"—were "ill."

Other new financial opportunities that arose from the publication of *DSM-III* were easy to see. The APA, with its insistence that the 265 disorders were diseases, had asserted a new authority over research into those ailments. If psychiatric problems were psychological in nature, then psychologists and other nonphysicians could easily compete with psychiatrists for grants from the National Institute of Mental Health to study such difficulties. But if such problems were discrete diseases, then research would focus on identifying their underlying pathologies and on developing treatments for the symptoms of these diseases, which was research that physicians could be expected to lead.

Finally, the new manual provided pharmaceutical companies with the opportunity to develop new drug treatments, and that, in turn, could be expected to benefit the psychiatric profession. The 1962 Kefauver-Harris Amendments to the Food, Drugs, and Cosmetics Act required that pharmaceutical companies prove that their new drugs were effective for *specific disorders*. Whereas no drug company could market a drug for "neurosis," which was seen as a psychological problem, it could market a drug for panic disorder, or posttraumatic stress disorder, or any of the other 263 disorders listed in *DSM-III*, now that they had been conceptualized as discrete illnesses. Academic psychiatrists could be involved in conducting the trials of the new drugs, and once they came to market, the field would have new products that could be expected to bring new patients to their offices.

A Seed for Corruption

With the light of hindsight, it is easy today to see the ethical peril for the APA that arose with its publication of *DSM-III*. The APA had devised a new manual that helped remake its image, and in a way that promised to benefit it in the marketplace, and the peril was that guild interests might now affect the story it told to the public about the nature of mental disorders, and the efficacy of somatic treatments for them. Science might have one story to tell, and yet the APA, because of guild interests, would have a need to tell another. That conflict, if it were not resolved in a manner that honored the science, had the potential to lead psychiatry seriously astray.

CHAPTER THREE

Economies of Influence

> Dependence upon funders produces a subtle, understated, camou-flaged bending to keep the funders happy...they become in the words of the X Files, "shape shifters" as they constantly adjust their views.
>
> —Lawrence Lessig, 2013[1]

As the Edmond J. Safra Center for Ethics has developed its lens for study-ing institutional corruption, it has focused on the need for researchers to identify the "improper dependencies"—or, alternatively described, the "economies of influence"—that may have proven corrupting to the insti-tution. In the previous chapter, we looked at how, with the publication of *DSM-III*, a potential conflict arose between the APA's guild interests and its ethical obligations as a medical discipline. The APA had a guild interest in promoting its new disease model, and yet, at the same time, it had an ethical obligation to inform the public, in a thorough and balanced way, what research was revealing about the nature of psychiatric disorders and the drugs used to treat them. In this chapter, we begin our inquiry into the conduct of the APA and organized psychiatry since that seminal moment in 1980, and, as a first step, it is necessary to detail the rise and growth of "economies of influence" that may have shaped psychiatry's behavior during the past 35 years. In this way, it is possible to see the financial influences that have been present during this period, and may have affected psychiatry's conduct.

There are three specific economies of influence to review: the APA's own guild interests, the influence of pharmaceutical money on the APA, and the influence of pharmaceutical money on academic psychiatry.

The APA's Guild Interests

With the publication of *DSM-III*, the APA, as a guild, had dominion over three "products" in the medical marketplace: diagnosis of mental disorders, research into their causes, and the prescribing of psychiatric

medications. From a commercial perspective, the APA had a need to promote this new disease model to the public. Thus, it is necessary to see
whether the APA, in its own reports since 1980, sought to do just that.
Did it, in essence, develop a public relations campaign designed to further
its guild interests? Did it make this type of campaign a regular feature of
its operations? If so, it will tell of a powerful "economy of influence"—the
promotion of guild interests—that has been present since 1980.

During the 1970s, there had been a great deal of internal dissension
within the APA, as the three factions—psychoanalytically oriented psychiatrists, social psychiatrists, and those with a biological orientation—
argued about different conceptions of mental illness, and how best to treat
such problems. With the publication of *DSM-III*, it was clear that the
biological faction had emerged triumphant in this battle, and the APA's
leadership quickly told its members that they needed to put past philosophical differences aside. "It is time to state forcefully that the identify
crisis is over," said APA medical director Melvin Sabshin.[2] In a 1981 editorial, the *American Journal of Psychiatry* explained what this meant, urging
APA members to "speak with a united voice, not only to secure support,
but to buttress [psychiatry's] position against the numerous other mental
health professionals seeking patients and prestige."[3] The journal, with this
editorial, was seeking to minimize dissent that might lead the public to
question the APA's new disease model of mental disorders.

With this goal in sight, the APA ramped up its capacity to tell—to the
scientific community and to the public—a story that, in fact, would promote its guild interests. The APA, explained Sabshin, in his 1981 report,
"is prepared to become much more active in efforts to educate the public
about the realities of psychiatry and to dispel some of the myths about our
field."[4] That year, the APA established a new "Division of Publications
and Marketing" that would, among other things, publish materials that
improved "public perceptions of psychiatry," which was part of an effort
to "deepen the medical identification of psychiatrists," Sabshin wrote.[5]
The APA also strengthened its public affairs department, which was
needed, said APA speaker William Sorum, "to improve our own public image and relationship with our constituent supporters...we must
represent ourselves as rational, scientific, and responsible."[6] In addition,
shortly after *DSM-III* was published, the APA launched a publishing venture, American Psychiatric Press, that would publish scientific material—
books, textbooks, and the *American Journal of Psychiatry*)—expected to
"have a significant impact on public opinion," Sabshin said.[7]

The APA's courting of the media, in this new *DSM-III* era, began
immediately. In December 1980, it held a daylong "National Media
Conference" in Washington D.C., which focused on "new advances in
psychiatry," and, as Sabshin happily noted, was "attended by representatives of some of the nation's most prestigious and widely circulated newspapers."[8] In 1982, it held two such seminars, including one in New York
City where "special attention" was given "to radio and television network

writers and producers, free-lance writers, and magazine editors."[9] The APA developed "Fact Sheets" for distribution to the press, which told of the prevalence of major mental illnesses listed in *DSM-III* and the effectiveness of psychiatric treatments. This multipronged media campaign proved successful, with the APA's public affairs division reporting a "fourfold increase" in contacts with the media by 1983.[10]

In his annual reports, Sabshin regularly celebrated the feature articles that were now appearing in the media. "With the help and urging of the Division of Public Affairs," he wrote in 1983, "*U.S. News and World Report* published a major cover story on depression, which included substantial quotes from prominent psychiatrists, among them APA President H.K. Brodie, M.D."[11] Three years later, the APA fielded "632 major media inquires" during the year, and, Sabshin reported, "APA spokespersons were placed on the Phil Donahue program, *Nightline*, and other network programs."[12] As media outlets reported on this "new psychiatry," the APA established a "media awards" program, honoring reporters with "letters of special commendation." In 1983–1984, David Zinman of *Newsday* received one for his article "The Fragile Mind." Barbara Walters also received an APA award that year.[13]

To better prepare its members to speak to the media, the APA established a "Public Affairs Network," which sought to develop a nationwide roster of "expert" psychiatrists who had been trained "in techniques for dealing with radio and television media." In 1985, the network sponsored nine "How to Survive a Television Interview" workshops around the country, with the APA reporting that by the end of that year, more than 300 psychiatrists had been schooled in this way. "We now have an experienced network of able leaders who can cope more effectively with all varieties of media," Sabshin said.[14] The APA taped radio interviews with many of those who had gone through the media training, with the interviews provided to local stations.

The APA also communicated this story of advances in clinical care directly to the public. The organization placed 90-second "public service spots" on cable television that were designed "to improve the image of psychiatry, among other things."[15] Harvey Rubin, chair of the APA's public affairs committee, recorded a popular radio program that reached "listeners around the country."[16] The APA encouraged district offices to conduct such campaigns, and provided them with "materials for use in newspaper columns," which was done to "ensure that appropriate information about psychiatry appears in the local news media."[17]

The APA also had good news to report about the impact of the association's new press. In 1982, Sabshin wrote that American Psychiatric Press had under contract a number of "major books targeted at the general public," such as *Psychiatric Illness: A Guide for the Family, 1001 Questions about Psychiatry,* and *Don't Panic,* and that it was "actively seeking new titles to publish, with an emphasis on material that will educate the American public on the positive aspects of psychiatry."[18] In 1985, Sabshin noted that the press "has been quite successful in promoting APA's educational and fiscal objectives."[19]

On the political front, the APA regularly invited the nation's political leaders to seminars on Capitol Hill, which were titled "Advances in Research and Clinical Care." The 1990 seminar, Sabshin reported, drew a "standing-room-only crowd of Congresspersons, staff, and community representatives."[20] This courting of Congress paid off, as the APA, in the mid-1980s, successfully lobbied the House and Senate to introduce joint resolutions proclaiming October 6–12 as "Mental Illness Awareness Week."[21] The Senate, in its resolution, declared that the "education and information disseminated through the week of its observance will help people understand that they do not have to suffer from debilitating anxiety, panic, phobias, depression, or schizophrenia."[22] Psychiatry, according to this Senate resolution, had developed very effective treatments for these illnesses, the drugs enabling people to live normal lives.

The APA, in these first post-*DSM-III* years, had implemented a comprehensive plan for promoting its disease model and improving public perceptions of psychiatry. It had reached out to the media, trained its members in doing interviews, developed "Fact Sheets," and successfully gotten Congress to promote its message. Once Upjohn began marketing Xanax as a treatment for panic disorder, and Eli Lilly brought Prozac to market, with these two events occurring in the late 1980s, the APA increased its public affairs efforts.

The APA developed both a two-year and five-year PR plan. "For the remainder of 1988 and 1989," explained APA secretary Elissa Benedek, "there will be an expansion of physician, media, member and public information activities under the campaign theme of 'Let's Talk About Mental Illness.'" That campaign would be the kick-off to a five-year effort that would include development of new fact sheets about mental illnesses "for distribution to news media and opinion makers"; the publication of a "periodic newsletter on mental illness topics for producers and writers for the entertainment media"; and "an expanded speakers bureau."[23] Public affairs, Sabshin noted in his 1987 report, was one of the APA's "highest priority areas," critical for "producing genuine advances in APA's objectives."[24]

At this time, with Upjohn eager to market Xanax as a treatment for panic disorder, and Eli Lilly touting Prozac as a breakthrough antidepressant, there was a noticeable shift in the focus of the APA's publicity efforts. Up until then, the APA had sought to broadly improve psychiatry's image, deepening its identification as a medical specialty and informing the public of "advances in clinical care." But starting in the late 1980s, the APA focused more on educating the public about anxiety and depression, and the effectiveness of treatments—that is, Xanax and Prozac—for them.

With funding from Upjohn, the APA created three "public education" films about the two disorders. *Panic Prison* hit the screen in 1989, while the second, *Faces of Anxiety*, premiered on May 2, 1990, at the Kennedy

Center in the nation's capital. In 1991, the APA premiered the third film, *Depression: The Storm Within*, at the same prestigious venue. All three films, Sabshin reported, were "extremely well received by viewers—psychiatrists, other health and mental health professionals, and laypersons." Nearly 1.5 million people saw the three films, which, Sabshin said, "have been designed to convey such messages as mental illnesses are real illnesses that cause pain and disability to millions of Americans; they can strike anyone; and they can be diagnosed accurately and treated effectively. The messages also clarify who psychiatrists are and what they do, as well as emphasize the impact of scientific advances."[25]

The APA, again with funding from Upjohn, also sponsored workshops on panic disorder. By the end of 1992, the APA announced that it had "reached over 10,000 physicians and other health professionals" through the workshops.[26] In a similar vein, the APA, this time with funding from Boots Pharmaceuticals, developed an "interactive workshop" on depression. This program, reported Benedek, proved to "be an effective means of reaching primary care physicians with a positive message about psychiatry."[27]

As other selective serotonin reuptake inhibitor (SSRI) antidepressants arrived on the market in the early 1990s, the APA joined with the National Mental Health Association to create a nationwide campaign to inform the public about depression and how it often went undiagnosed. This campaign, which featured "major national advertising coverage," had three main messages: "clinical depression is a medical illness, effective treatments are available, and see a doctor for help." In each major market, the APA identified local psychiatrists who could serve "as medical experts and public and media spokespersons." The campaign was first rolled out in Jacksonville, Florida, and it prompted more than 2,000 people to call a hotline to request free materials on depression, with area doctors reporting "an increase in patients seeking treatment."[28]

At this same time, the APA, as part of its Mental Illness Awareness Week, began sponsoring a "National Depression Screening Day." The number of people screened grew dramatically each year; in 1994, Sabshin wrote that an estimated 65,000 people had been screened at 1,300 sites across the country. The theme of Mental Illness Week that year was "Treatment Works!", a comforting message for those newly identified as depressed during the screening process.[29]

These two initial educational efforts, focused on anxiety disorders and depression, were soon complemented by a more general "Let's Talk About Mental Illness Campaign" that the APA mounted in the early 1990s. In one of his annual reports, Sabshin announced that the APA had distributed "almost a million products," in the form of "brochures, pamphlets and other written materials" on mental disorders and "choosing a psychiatrist." It created a "Let's Talk About It" comic book "designed for "junior and high school audiences"; developed a "Let's Talk About Mental Illness" film series that Blockbuster video stores carried; worked with the marketing department of *Reader's Digest* to create "mental illness information columns" for the popular magazine; and set up meetings with the editorial boards and staff at a number of newspapers, "including the *Wall Street Journal, San Francisco Examiner, San Francisco Chronicle* and *Washington Post*."[30] This public affairs effort, the APA reported, had "led to excellent television, radio, and print media coverage."[31]

By the end of the 1990s, the APA could look back and see a 20-year record of public relations successes, with its efforts having reshaped public understanding of psychiatric illnesses and its treatments. "Focus groups indicate that members of the general public increasingly reflect the attitude that mental illnesses are diseases with effective treatments," the APA wrote in one of its annual reports. "We are dispensing a powerful message to the general public and they are hearing it...The APA has entered the 21st century as more than a professional organization. It has emerged as a pertinent, extremely effective provider of essential mental health information to the public."[32]

Since that time, the APA has regularly continued such efforts, initiating one campaign after another. In 2005, it initiated an effort titled "Healthy Minds, Healthy Lives," which focused on delivering the message that "mental illnesses are real and treatable," and was designed to "highlight the importance of seeking treatment from psychiatric physicians."[33,34] The following year, it joined with six other organizations to launch a "Depression is Real" campaign, which promoted the understanding that "depression is a serious, debilitating disease that can be fatal if left untreated."[35] To deliver a similar message about pediatric depression, the APA created a website, ParentsMedGuide.org, and it organized a campaign titled "Typical or Troubled" that urged schoolteachers to be on the lookout for children with signs of mental illness and to refer them for help.[36] In 2009, it strengthened its organizational capacity to deliver such messages, developing a new Council on Communications that was charged with "creating excitement about psychiatrists' ability to prevent and treat mental illness; and branding psychiatrists as the mental health and physician specialists with the most knowledge, training and experience in the field."[37] The APA also developed a social media campaign, using Twitter and Facebook to promote its "treatment works" message, with the APA reporting in 2012 that its Facebook page now had 18,150 "likes."[38]

Such have been the APA's efforts since the publication of *DSM-III* to tell the public a story designed, as its leaders acknowledged, to "advance the profession of psychiatry." The APA urged its members to speak with a united voice, provided media training to its members who could serve as spokespersons at the local and national levels, and launched a press that was expected to publish scientific books, texts, and journals that would positively impact public opinion. The APA mounted one publicity campaign after another, and the message it promoted was that psychiatry was making great advances in clinical care; that mental disorders were real diseases; that its treatments worked; and that such diseases, which were common in children and adults, often went untreated.

In short, the APA, in its communications to the public, clearly told a story meant to advance guild interests. Now if the story that the APA told was, in fact, consonant with scientific findings, there would be no harm done. But if guild interests were prompting the APA to tell a story unmoored from science, such that, in its details, the story could be seen as incomplete or, at times, at odds with research findings, then that would be evidence of a profession that had been corrupted by those guild interests.

Partnering with Industry

While the influence of the pharmaceutical industry on the APA became much more pronounced following the publication of *DSM-III*, money from drug companies was already flowing to the APA before that seminal event. Chlorpromazine was introduced in 1954, initiating what is often called today a "psychopharmacological revolution." Over the next decade, as pharmaceutical companies brought new psychiatric drugs to market, they flooded the *American Journal of Psychiatry* and other psychiatric journals with advertisements. Industry money was now bolstering the APA, and to such an extent that in 1974 its board members worried that "APA's relationships with pharmaceutical companies were going beyond the bounds of professionalism [and] were compromising our principles."[39]

In response to that concern, the APA formed a task force to study what the impact would be on its operations if it lost such pharmaceutical support. How dependent had the APA become on industry? The task force concluded that, without industry money, many local APA organizations and various training programs would fold. Faced with this conclusion, the APA, rather than seek to curtail its relationships with industry, sought instead to expand them, deciding that the "obtaining of grants and contracts to support needed programs" should be a "top priority." This decision quickly paid off for the APA, as the grants it received rose from $294,842 in 1976 to $1.36 million in 1980.[40]

With the APA's adoption of a disease model in 1980, it provided industry with a powerful new reason to court the profession with its dollars. Since the APA was now declaring that mental disorders were brain diseases, with hundreds of such "diseases" listed in *DSM-III*, pharmaceutical companies could market drugs for these various illnesses, with drugs expected to be the first-line treatment for many of them. Drugs, after all, were approved because they were effective against the "symptoms" of a disease. While this was not a result that motivated Spitzer's remaking of the DSM, the new manual nevertheless provided pharmaceutical companies with a way to dramatically expand the market for their products. "The pharmaceuticals were delighted" with *DSM-III*, Spitzer acknowledged years later.[41]

However, the APA did embrace the financial opportunity that was now present. In 1980, its board of directors "voted to encourage pharmaceutical companies to support scientific or cultural activities rather than strictly social activities as part of the annual meeting program."[42] The APA began allowing pharmaceutical companies to sponsor scientific symposia at its annual meeting, which took the form of presentations given by academic medical school faculty at breakfast, lunch, and dinner meetings. In addition to paying the APA a fee for this opportunity, the drug firms would place advertisements in the APA's journals months before the annual meeting to promote the upcoming mealtime events.

Drug firms jumped at the chance to sponsor such events. In 1982, five companies sponsored a mealtime symposium, and they placed eight pages of advertising in the APA's journals. This was the beginning of a deluge, and by the late 1990s, the APA was reporting that there were more than 40 such events each year. At the 2004 annual conference, there were 54 industry-supported symposia, with the pharmaceutical companies reportedly paying $50,000 for each symposium.[43] Collectively they paid for 261 pages of advertising in APA journals that year to tell of their upcoming mealtime events. "Advertising revenue," wrote medical director James Scully in 2004, "helps APA meet its objectives."[44]

At the annual meetings, pharmaceutical companies also paid for lavish exhibits in the conference exhibition hall. Psychiatrists could wander through the exhibits, hearing about the wonders of psychiatric drugs and collecting various trinkets, gifts, or, as was reported one year, lying down for a free massage or sitting to have their portrait painted. "The enormous exhibit area at the modern convention center also serves as a colorful image of the entire Annual Meeting," Sabshin marveled in his memoir. "A new 'city' has been quickly constructed in the exhibit hall—one that is dominated by pharmaceutical companies. New psychiatric medications are featured, but old 'favorites' continue to get some attention."[45]

With the pharmaceutical companies paying for the symposiums, filling the exhibit hall with their exhibits, and funding social activities, the annual meeting became the APA's most profitable activity. Meeting revenues rose from $1 million in 1980 to $3.1 million in 1990, and then revenue tripled again in the 1990s, reaching $11.3 million in 2000. Four years later, the APA's meeting revenues hit $16.9 million, producing a profit of $9.8 million for the APA that year.[46]

However, commercial influence on the APA's annual meeting was only the most visible part of a much larger story. After *DSM-III* was published, the drug companies became ever more interested in supporting APA programs. The APA found that it could turn to pharmaceutical companies for money to fund its public affairs campaigns, its political lobbying efforts, and other special projects. Pharmaceutical companies could also be tapped to pay for fellowships for young psychiatrists.

As noted earlier in this chapter, after *DSM-III* was published, the APA organized media workshops to train psychiatrists around the country on how to speak to the media. Pharmaceutical companies helped fund the workshops. In 1982, when the APA launched a political action committee, the committee was supported by pharmaceutical money.[47] That same year, Mead Johnson Pharmaceuticals gave the APA $50,000 for a fellowship.[48] And so forth: in 1985, a drug company provided funding to support subscriptions to one of the APA publications.[49] In 1987, when the APA sponsored a conference on how psychiatry could secure more influence, it was underwritten by several drug companies, including Upjohn.[50]

In the late 1980s and 1990s, the APA's collaborations with industry deepened. Drug companies were bringing new products to market, and they provided funding to the APA to mount campaigns to "educate" the public about depression and anxiety. One of the first such campaigns was funded by Boots Pharmaceuticals, which sponsored 100 workshops on depression around the country. These efforts, announced APA secretary Elissa Benedek in 1988, had "proven to be an effective means of reaching primary care physicians with a positive message about psychiatry."[51]

The APA, in its annual reports, celebrated this joint public affairs effort with industry. "We have been successful in our public affairs functions," noted the APA in 1988. "Collaborative projects with industry have helped us enhance psychiatry's image with other physicians and the public."[52] This theme was echoed again and again, with the APA reports noting that such public information campaigns were made possible by funds from industry.

Industry's sponsorship of APA fellowships, which was first noted in the APA's 1983 report, grew in number as the years passed. By the early 2000s, there was the APA/Lilly Chiefs Residents leadership program, the APA Bristol Myers Squibb Fellowship in Public Psychiatry, the APA Glaxo Wellcome Fellowship, the SmithKline Beecham Young Faculty Award, the APA Lilly Psychiatry Research Fellowship, and the APA Wyeth-Ayerst M.D./Ph.D. Psychiatric Research Fellowship.[53]

When the APA's annual reports are carefully read, they occasionally tell of industry providing the funds for even the most routine APA functions. For instance, in 1992, when the APA was looking to fill three senior positions, including a director of the office of education, the search process was funded by Abbott Laboratories. James Scully was hired for the education post, with his salary paid by Abbott. "We do not see this important position as solely funded by an outside firm," Sabshin wrote in 1992. "Rather this donation enables us, particularly in these tight fiscal times, to recruit a psychiatrist director who will bring considerable experience and involvement to the position. We had delayed filling the position because of fiscal constraints."[54]

This pharmaceutical funding, which flowed into the APA in so many ways, helped transform the APA from—as Sabshin put it—a "Mom and Pop" organization into a "business-like mammoth organization."[55] In 1987, the APA moved into a fancy new building in Washington D.C., and nearly every year, the APA reported rising revenues. Annual revenues

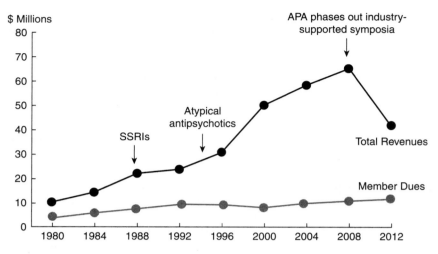

Figure 3.1 APA's annual revenues, 1980–2012.

Source: APA's annual financial reports, 1980-2012.

increased from $10.5 million in 1980 to $24.9 million in 1990, and then doubled again in the 1990s, hitting $50.2 million in 2000. The APA's revenues topped out at $65.3 million in 2008 (Figure 3.1).[56]

It is impossible to identify the precise percentage of the APA's annual revenues that came directly from the pharmaceutical industry during this period (from 1980). On several occasions in its annual reports, the APA did provide a number: 29 percent of annual revenues came from industry in 2000; an equal percentage in 2006; and then it dropped to 21 percent in 2008.[57] However, those figures do not capture the *indirect* ways that industry money flowed to the APA, or supported its activities. For instance, the pharmaceutical companies regularly paid the expenses of psychiatrists coming from abroad to attend the APA's annual meeting, and while this benefitted the APA, as it increased the number of people attending the conference and paying the conference fee, the travel money from the pharmaceutical companies went to the individual psychiatrists, and thus it wouldn't show up on the APA's books as a direct grant from industry.[58]

In addition, in 1991 the APA created a 501(c)(3) nonprofit affiliate, the American Psychiatric Foundation, which received grants from pharmaceutical companies and other donors to fund "public outreach" programs. The American Psychiatric Foundation helped fund the National Depression Screening programs organized by the APA in the 1990s; a decade later, it funded the national distribution of the Healthy Minds TV shows to PBS stations around the country.[59] While the APA, in its annual financial reports, did not regularly detail the amount that the American Psychiatric Foundation had received, treasurer Carol Bernstein did report that in 2002 it had raised $2.6 million for "private grant support" of the

APA and its activities.[60] Much of this funding came from the pharmaceutical industry; for instance in 2008, when four of the 15 members of the foundation's board were "high-level executives from pharmaceutical companies," 11 drug firms gave money to the foundation.[61]

In 1998, the APA created a second 501(c)(3), the American Psychiatric Institute for Research and Education. Its purpose was to focus on "research" enterprises, and this subsidiary of the APA was also heavily supported by industry. In 2008, nine of the board's 16 members had industry ties.[62]

All of this tells of how an "economy of influence" developed within the APA following the publication of *DSM-III*, with the amount of funds flowing from industry to the APA, through one channel or another, increasing steadily for the next 25 years. Very early on, Speaker Fred Gottlieb, in his 1985 report, warned about the potential for this money to corrupt the organization:

> In May 1984, I wrote about the *millions* of dollars of drug house money we receive annually from advertising and commercial exhibits and for awarded lectureships and social functions. I was surprised to learn, when I attended the APA public affairs workshop in Tucson, that pharmaceutical money underwrites a portion of such activities. I was startled to learn that our Federal Legislative Institute is similarly supported.
>
> I do not suggest that either they or we are evil folks. But I continue to believe that accepting such money is, in the long run, inimical to our independent functioning. We have evolved a somewhat casual and quite cordial relationship with the drug houses, taking their money readily because it helps promote events that we consider worthwhile and feel we could not otherwise afford. We seem to discount available data that drug advertising promotes irrational prescribing practices. We seem to think that we as psychiatrists are immune from the kinds of unconscious emotional bias in favor of those who are overtly friendly toward us. We apparently assume we are too wise to suppress cognitively dissonant information. We persist in ignoring an inherent conflict of interest.[63]

Gottlieb, when raising this concern, noted that a previous APA program committee had worried "about the problem of our being, either in fact or just in perception, significantly controlled by economic forces external to the organization, thereby adversely affecting the public perception of the psychiatric profession as being independent from undue influence by corporate interests with their own manifold needs and priorities."[64] Yet, in spite of Gottlieb's alarm, the APA board didn't attempt to rein in this influence. Instead, many in the APA now told of how the organization and industry had developed a "partnership," a relationship captured by a photo in the January 2, 1987, issue of *Psychiatric News*, which showed

APA president Robert Pasnau receiving a check from Smith, Kline, and French.[65]

During the 1990s, a growing number of psychiatrists questioned whether this "partnership" was ethical. For instance, in a 1990 article, Matthew Dumont publicly complained that "psychiatry and the pharmaceutical industry" were now "in bed together."[66] In 1998, Loren Mosher, former head of schizophrenia studies at the NIMH, resigned from the APA, calling it the "American Psychopharmacological Association."[67] The criticism grew louder, and in 2007, psychiatrist Daniel Carlat told the *Boston Globe* that "our field as a whole is progressively being purchased lock, stock, and barrel by the drug companies: this includes the diagnoses, the treatment guidelines, and the national meetings."[68]

With newspapers also starting to write about this conflict of interest, the APA began to worry that it was damaging the profession's reputation. In 2002, APA president Paul Appelbaum expressed his dismay over the most visible example of this conflict: the gift grabbing by psychiatrists in the exhibit hall at the annual meeting each year.

> Are there any psychiatrists in America who can't afford to buy their own pens, notepads, and Frisbees or to call home from the annual meeting on their own dime? What other than a feeding frenzy can explain the shopping bags crammed with giveaways that some attendees can be seen lugging from the convention center. Do they really send their kids to the park to play with Frisbees emblazoned with the name of a popular SSRI? I don't blame the companies for trying to foist this stuff on us; I blame us for taking it.[69]

A short while later, Appelbaum focused on the larger issue at stake. "Could APA exist without any money coming from the pharmaceutical industry?" he asked its members. "Of course it could; [but] it would be a much smaller organization, and the tasks it could undertake would be much more limited."[70]

The breaking point for the APA, in terms of having to worry about its reputation, came in 2008, when Iowa Senator Charles Grassley, who had been exposing some of the large sums that industry was paying individual psychiatrists, asked the APA for a "complete accounting of [its] revenues since 2003, excepting those from advertising in journals."[71] The senator noted that "I have come to understand that money from the pharmaceutical industry can shape the practices of nonprofit organizations that purport to be independent in their viewpoints and actions."[72]

In response, the APA formed a "Work Group on Future Relationships With Industry" to address the fundamental question of "whether APA should have or maintain relationships with industry."[73] Although it concluded that cutting all ties wasn't "feasible," the APA did take one notable step to limit this influence, deciding in March 2009 that it would phase out industry-supported symposia at its annual meeting, reversing the very policy that, since its enactment in 1980, had proven so profitable for the APA. This decision coincided with the fact that drug companies were no longer bringing many new psychiatric drugs to market, which meant that the industry had much less reason to place advertisements in the APA's journals.[74]

"My friends, in terms of revenues today, we are much smaller than we were two years ago," APA president Alan Schatzberg told the audience in his 2010 presidential address. "Our four corporations now have total annual revenues of around $55 million, down from $65 million a few years ago. This is due largely to the loss of ad revenues in our journals." There were, he added, almost no industry-supported symposiums at its annual meeting that year. "This is not only of our doing, but the negative attacks on industry have made them gun shy of supporting such programs. That is too bad, but that is where the field is heading."[75]

This "economy of influence" over the APA, while not disappearing altogether, was on the wane. Meeting revenues declined; the APA's total revenues dropped from $65.3 million in 2008 to $42.6 million in 2012. That year, the APA reported receiving only $5.6 million from pharmaceutical companies, less than half the reported amount from industry five years earlier.[76] The partnership is no longer what it had once been, when, for nearly three decades, industry regularly provided the funding for the APA's "educational efforts" and numerous other "special programs," and the revenues from industry at its annual meeting produced a profit of millions of dollars for the APA.

Industry's Thought Leaders

Up until the 1980s, academic physicians generally remained aloof from industry. Grants from the National Institute of Health were the coin of the realm for academic researchers, and drug companies often told of how, when they sought to get academic physicians to conduct clinical trials of their drugs, they had to approach the academic researchers with "hat in hand." This aloofness began to thaw in 1980 once Congress passed the Bayh-Dole Act. Prior to that time, discoveries made by NIH-funded researchers remained in the public domain, and any company could exploit them. Bayh-Dole allowed academic researchers who had made NIH-funded discoveries to license their discoveries to drug companies and collect royalties. The individual researcher and his or her academic institution could now profit from an NIH-funded discovery, and thus academic medicine had a new reason to collaborate with industry. This impulse became even more pronounced in the subsequent years when it grew more difficult for academic researchers to obtain NIH grants.

The Bayh-Dole act was passed the same year that the American Psychiatric Association decided to allow pharmaceutical companies to sponsor scientific symposiums. That decision opened the door for drug companies to pay academic psychiatrists to serve as their speakers, which provided industry with a new influence over the doctors. As Sabshin acknowledged, "these symposia are meticulously prepared with rehearsals before the meeting.[77]

Once pharmaceutical companies began hiring academic psychiatrists to be their speakers, they also began paying them to serve as advisers and consultants. Industry insiders came to refer to these experts, in psychiatry and other fields, as key opinion leaders (KOLs) or "thought leaders." They could be involved in every aspect of the drug-development process: providing advice on how to design the trials; serving as investigators in the clinical trials; authoring papers that told of study results; talking to the media after those results were published; and then, following the FDA's approval of the drug, speaking about it and the related disorder at CME conferences and dinners organized by the pharmaceutical company. The ubiquitous nature of such ties between industry and psychiatrists can be seen in the disclosure reports filed by 273 speakers at the APA's annual meeting in 2008. Collectively, they told of having 888 consulting contracts and 483 contracts to serve on speakers' bureaus.[78]

As a result, this financial conflict of interest came to pervade every corner of the psychiatric enterprise. Editors of journals had such ties to industry; so too did writers of psychiatric textbooks; authors of clinical practice guidelines; developers of diagnostic criteria; and chairs of psychiatric departments at medical schools. For instance:

- In the early 1990s, when *DSM-IV* was created, 57 percent of the task force members reported having ties to industry; all of the panelists for the mood disorders and schizophrenia work groups had ties to industry.[79] Fifteen years later, when the APA was creating the fifth edition of this manual, 69 percent of the task force members reported they received funds from industry.[80]
- Of the 20 work group members who, prior to 2009, authored the APA's guidelines for depression, bipolar illness, and schizophrenia, 18 (90 percent) had financial ties to industry.[81]

The reward for academic psychiatrists who became industry thought leaders went beyond money. A KOL was presented to the public as a leader in the field, which in turn brought new career opportunities (and of course was flattering to the ego). "These experts get not only money and visibility, but power, particularly if they become members of special interest groups," said Giovanni Fava, in a 2008 article in the *British Medical Journal*. "Because of the resultant contacts, members of these groups often get leading roles in editing medical journals, advise nonprofit research organizations, and act as reviewers and consultants."[82]

From the standpoint of institutional corruption, it is not important to assess whether this industry money did, in fact, alter the behavior of *individual* academic psychiatrists in ways that that we would see as "corrupt."

Table 3.1 Examples of pharmaceutical payments to thought leaders in psychiatry

Academic Psychiatrist	Affiliation	Pharmaceutical Company	Amount
Joseph Biederman	Professor at Harvard Medical School	Janssen	$1.6 million (2000–2007)
Frederick Goodwin	Former NIMH Director	GlaxoSmithKline	$1.2 million (2000–2008)
Melissa DelBello	Associate Professor at Univ. of Cincinnati	Astra Zeneca	$418,000 (2003–2007)
Karen Wagner	Director of Child Psychiatry at Univ. of Texas	GlaxoSmithKline	$160,000 (2000–2005)

Source: Senator Charles Grassley, "Disclosure of Drug Company Payments to Doctors," 2008.

Instead, as we noted in our introduction, we are presuming, as Lawrence Lessig said, that individuals within the institution are innocent, but that the "economy of influence" that they allowed to develop is not. Thus, we simply needed to detail how this particular "economy of influence" developed, and how pervasive it came to be. Academic psychiatrists are the experts that inform both society and their peers about the validity of psychiatric disorders, and the safety and efficacy of drug treatments. They also establish diagnostic criteria and develop clinical care guidelines. What we see in this quick review is that during the past 30 years, academic psychiatry performed this task while a majority of its members were being paid by pharmaceutical companies to serve as advisors, consultants, and speakers (Table 3.1).

A Perfect Storm

At the end of the previous chapter, we described the "ethical peril" that arose when the APA, with its publication of *DSM-III*, adopted a "disease model." The peril was that the APA's guild interests might trump its fidelity to its scientific obligations. And what we see in this chapter is that since 1980 the APA has regularly conducted media campaigns designed to promote guild interests; that those media campaigns were at least partly funded by pharmaceutical companies; that such industry money helped the APA grow from a trade organization with revenues of $10 million in 1980 to one with $65 million in revenues in 2008; and that, starting in the mid-1980s, pharmaceutical companies began paying academic psychiatrists to serve as their "key opinion leaders." All three of these "economies of influence" pushed psychiatry in the same direction, which was for the field to tell of validated diagnoses, advances in research, and drug treatments that were highly effective.

Now, in the next part of this book, we will examine the behavior of organized psychiatry during this period.

PART II

Science Corrupted

CHAPTER FOUR

The Etiology of Mental Illness Is Now Known

Health care workers are obliged to use the best medical science to relieve suffering and pain, treat illness, and address risks to health. The institutional corruption of health care consists of deviations from these principles.

—Donald Light, 2013[1]

It is easy to detail our society's expectations of a medical specialty. We expect that the field will put the interests of patients—and thus of society—first. This public interest should trump the field's interests as a guild. As such, we expect that the leaders of the specialty will provide the public with accurate information about the nature of the disorders they treat, and about the risks and benefits of treatments. We also expect that the profession will conduct research designed to improve the care it is able to provide, and that it will follow the dictates of good science as it pursues that goal.

This set of expectations provides us with a foil for assessing the behavior of the American Psychiatric Association (APA)—and organized psychiatry—since 1980. We will want to see whether the APA and the leaders in the field have been trustworthy in their communications to the public, and whether, as they assessed the merits of their diagnostic manuals and of their treatments for psychiatric disorders, they have done so in a manner consistent with good science. We can start by reviewing the APA's introduction of *DSM-III*, and its subsequent descriptions of the reliability and validity of its diagnoses.

"The Reliability Problem Is Solved"

In infectious medicine, a diagnostic manual needs to be both reliable and valid in order to be truly useful. A classification system that is reliable enables physicians to distinguish between different diseases, and to then prescribe a treatment specific to a disease, which has been validated—through studies of its clinical course and, if possible, an understanding of its pathology—as real. In 1978, psychiatrist Gerald Klerman, director of the Alcohol, Drug and Mental Health Administration, which housed the NIMH at that time, declared that psychiatric disorders should be seen in this same light. "There is a boundary between the normal and the sick. There are discrete mental illnesses…It is the task of scientific psychiatry, as a medical specialty, to investigate the causes, diagnosis, and treatment of these mental illnesses."[2]

As Spitzer and his colleagues developed *DSM-III*, they emphasized that it needed to provide the field with a way to reliably distinguish between different psychiatric conditions. Both *DSM-I* and *DSM-II* had failed to do that, and while it might take time for psychiatry to "validate" the disorders listed in *DSM-III*, the new manual would be rather useless if it did not provide diagnostic reliability. "A necessary constraint on the validity of a [nosological] system is its reliability," Spitzer wrote. "There is no guarantee that a reliable system is valid, but surely an unreliable system must be invalid."[3]

Even so, when the field trials of *DSM-III* were launched in 1977, Spitzer told the NIMH that "the major purpose of the study is to identify and solve potential problems with the *DSM-III* draft." This study was primarily a debugging exercise, with the clinicians expected to provide feedback to Spitzer so that the APA could amend the diagnostic criteria to improve reliability. While the APA would use a variety of clinical settings for the trials, Spitzer said that it was important that the trials were conducted by clinicians "with sufficient motivation to provide us with the kind of quality data that is required." As such, Spitzer informed the NIMH that "no attempt will be made to insure that the types of facilities and patients be statistically representative in this country."[4]

However, when Spitzer published results from the first phase of the trials in June 1979, he presented them in a different light. The trials, he declared, had proven the reliability of the new manual to be markedly superior to *DSM-I* and *DSM-II*. "The interrater agreement for major diagnostic categories in studies using *DSM-I* and *DSM-II* was usually only fair or poor. In phase one of the *DSM-III* field trials the overall kappa coefficient of agreement for axis 1 diagnoses of 281 patients was .78 for joint interviews and .66 for diagnosis made after separate interviews." These results, Spitzer wrote, showed that "for most of the [diagnostic] classes, the reliability…is quite good…these results were so much better than we had expected…It is particularly encouraging that the reliability for such categories as schizophrenia and major affective disorders is so high."[5]

This was a claim of a *scientific* success. A year later, in a paper published in the *American Journal of Psychiatry* that provided an overview of the making of *DSM-III*, Spitzer and his colleagues reiterated this same point, and they told of a robust scientific process at work:

The need for reliability, that is, agreement among clinicians on assigning diagnoses to patients, is universally acknowledged. Studies of the reliability of psychiatric diagnosis using *DSM I* and *DSM II* indicated generally poor or only fair reliability for most of the major diagnostic categories. In the *DSM III* field trials over 450 clinicians participated in the largest reliability study ever done, involving independent evaluations of over 800 patients—adults, adolescents, and children. For most of the diagnostic classes the reliability was quite good, and in general it was much higher than that previously achieved with *DSM I* and *DSM II*.[6]

DSM-III was published shortly thereafter, and in the introduction, Spitzer, once again, emphasized this finding. The most important aspect of the field trials, he said, "was the evaluation of diagnostic reliability," and that the results "generally indicate far greater reliability than had previously been obtained with *DSM II*."[7]

This was the bottom-line result. The new manual had been field tested, and it provided psychiatry with a way to diagnose mental disorders with much greater reliability than before. The picture being presented to the field and to the public was this: Psychiatrists working in different psychiatric settings, when faced with the same group of patients, would give the same diagnosis to most of the patients.

Psychiatry, it seemed, had taken a big step forward. It could now diagnose disorders with the same degree of reliability as other medical specialties could, and many in the APA hailed *DSM-III* as a remarkable scientific achievement. "The development of *DSM III* represents a fateful point in the history of the American psychiatric profession," Klerman declared. "Its use represents a reaffirmation on the part of American psychiatry of its medical identity and its commitment to scientific medicine."[8] The new manual, said APA president Donald Langsley, in a 1981 address, "offers increased standardization of diagnosis, permitting a more precise definition of disease and increased precision in understanding cause and treatment."[9]

While there were some within psychiatry who questioned *DSM-III*'s utility, particularly since the disorders had not been "validated," nearly everyone accepted Spitzer's conclusion that diagnostic reliability had been improved. At the APA's 1982 annual conference, Klerman bluntly declared that the reliability problem had been "solved," and argued that the APA had blazed a new trail in assuring this was so. "*DSM-III* underwent field testing for reliability. Never before have practitioners of a medical specialty participated in a test of the reliability of their nomenclature. Never before has statistical evidence been brought forth with respect to the acceptance, reliability, feasibility, and utility of a new diagnostic system."[10]

DSM-III was an instant success. In the first six months following its publication, the APA sold more copies of its new manual than it had previously sold of its two prior *DSM* editions combined.[11] *DSM-III* was quickly adopted by insurance companies, educational institutions, courts and governments at all levels, and it became the official nosology of researchers, first in the United States and then increasingly abroad, as it was translated into more than 20 languages. *DSM III* became psychiatry's new "Bible" throughout much of the world.

A Closer Look at the Field Trials

Within a framework of institutional corruption, the question that now needs to be investigated is whether Spitzer and the APA, as they touted the new manual's reliability, were telling a story shaped by good science. As was briefly noted in chapter 2, the field trials—in their design and methods—could not be said to have provided a rigorous scientific test of *DSM-III*'s reliability, and if we now examine the trials more closely, we can detail all the ways they fell short of that standard. Specifically:

- No effort was made to ensure that the clinicians involved were a representative sample of clinicians in the United States. Instead, Spitzer solicited the participation of psychiatrists through notices in *Psychiatric Times* and other journals. In other words, it was a self-selecting process.[12] Furthermore, some of those who had helped create the diagnostic categories participated as investigators in the field trials.
- The clinicians who were paired to assess whether their diagnoses would be the same were not randomly selected. Instead, as two academic psychiatrists who reviewed the trials noted, the paired clinicians were "close colleagues," and "each clinician chose his own partner in the study." There was no control over whether those close colleagues then consulted each other before making their diagnoses, which were supposed to be independently made.[13]
- No effort was made to ensure that the field trials enrolled a representative sample of patients in the United States. Instead, two clinicians at a site, after having gained some experience using the *DSM*, would select one to four patients to jointly interview, which obviously led to the possibility that the clinicians would choose patients with a diagnosis that was immediately evident.[14]
- Spitzer, as he solicited feedback from the clinicians, even raised concerns about these lack of controls. Had they selected a case "because of an obvious great diagnosis," he asked. "Did you speak to your partner about your diagnosis after the patient interview...Did you change your diagnosis to be in accord with your partner? Did you not send back a diagnosis because you and your partner disagreed?"[15] While Spitzer understood that these factors would undermine the credibility of the kappa scores, he did not publish any report stating that when the protocol was violated in these ways that the data was then excluded.

- Historian Hannah Decker, in her review of archival material, found that many of the clinicians did not read the extensive *DSM-III* text, but rather relied on criteria they knew well from *DSM-II*, and in this way "hoped that they could fudge their way into a *DSM-III* diagnosis."[16] In fact, 65 percent of the clinicians reported that they had spent less than ten hours familiarizing themselves with *DSM-III*, even though it listed 265 disorders.[17]
- Spitzer and his colleagues only published "selective" parts of the data they collected. No final summary of the data was ever published, and the APA did not allow other investigators to reanalyze the data from the field trials.[18]

Much of this critique was put forth by Stuart Kirk and Herb Kutchins in their 1992 book, *The Selling of DSM*. They concluded that these deficiencies were so pronounced that, "in fact, 'meaninglessness' may be close to what can be claimed about the reliability field trials."[19] Many years earlier, Michael Rutter and David Shaffer, both academic psychiatrists, had come to a similar conclusion. "As pieces of research [the field trials] leave much to be desired... the findings do little to provide a scientific basis for *DSM III*."[20]

Such were the methodological flaws in the field trials. But as Kirk and Kutchins detailed in their book, there was one other revealing fact to be gleaned from a careful parsing of Spitzer's reports. His claims that the kappa scores for *DSM-III* were much better than those for *DSM-I* and *DSM-II* relied, in large part, on linguistic sleight of hand.

In Spitzer's 1974 review of *DSM-I* and *DSM-II* reliability studies, he used kappa scores to assess the diagnostic agreement. On this scale, a score of zero meant that the diagnostic agreement between two physicians assessing the same patient was no better than chance, while a score of 1.0 meant that the two physicians were in perfect agreement. As Kirk and Hutchins detailed in their book, Spitzer, in his discussion of the scores, decided that a mean kappa score of .40 for a diagnostic category was "poor," a mean score of .56 was "no better than fair," and a score of .70 was "only satisfactory." As such, Spitzer concluded that kappa scores for the *DSM-I* and *DSM-II* manuals were "only satisfactory" for three diagnostic categories, "no better than fair for psychosis and schizophrenia," and "poor" for the remaining categories. No category earned a "uniformly high" score, which presumably would have required a kappa score above .80.[21]

However, when Spitzer reported the kappa scores from the *DSM-III* field trials, he revised his standards for evaluating the results. He wrote that a "high kappa (generally .7 and above) indicates good agreement as to whether or not the patient has a disorder within that diagnostic class." What had previously been "only satisfactory" was now "good." Furthermore, that was just the first step of a two-step process that enabled Spitzer to depict the *DSM-III* results in such a favorable light. In fact, only 31 of the 80 reported kappa scores for adults on axis I major mental disorders reached that .70 level, and thus more than 60 percent were properly characterized—even by Spitzer's lowered standards—as "only satisfactory" or worse. However, when Spitzer summed up the *DSM-III* results, he stated that "for most of the diagnostic classes the reliability was quite good."[22] His conclusion was out of sync with the kappa scores, even when judged by his revised standards for "good."

Not surprisingly, subsequent reliability studies of *DSM-III* in real-world settings did not duplicate the "good" results reported by Spitzer. In a study of 154 first episode patients in British Columbia, the overall kappa score was .21, far below what had been reported in the *DSM-III* field trials. "Agreement between researchers and clinicians on diagnoses was fair to poor," the Canadian investigators concluded.[23] A 1985 study of psychiatric patients coming to emergency rooms produced kappas ranging from .29 to .62 for major diagnostic categories, which led the investigators to conclude that the diagnosis of "specific subtypes of mental illness, such as schizophrenia and bipolar disorder, were not made reliably in the emergency room."[24] Most notable of all, the NIMH ran a study during the 1980s that assessed the reliability of diagnosis made for 810 patients in Baltimore. The researchers calculated kappa scores for eight major diagnostic categories, and none produced a kappa above .35 and seven of eight were below .26; these scores were far below those reported by Spitzer in his initial paper on the *DSM-III* field trials. None of the kappa scores, the NIMH researchers concluded, "suggests more than modest agreement beyond what is expected by chance."[25]

All of these follow-up studies led to the same conclusion. There was no good evidence that *DSM-III* provided the field with a "reliable" nosology, and in fact, there was no good evidence that reliability with *DSM-III* was better than with *DSM-I* or *DSM-II*. However, the story that was embraced by the APA and its leaders in the early 1980s was of a different sort, and in those circles, the understanding was that reliability with the new manual was good. And that belief was the principal "scientific" selling point for a manual that would dramatically alter societal thinking about the nature of psychiatric problems.

Validating the DSM

While the published articles on the field trials may have led to a general understanding that psychiatry's new manual provided diagnostic reliability, the APA, much to its credit, acknowledged in the manual's introduction that most of the 265 disorders had yet to be validated. "It should be understood, however, that for most of the categories the diagnostic criteria are based on clinical judgment, and have not yet been fully validated by data about such important correlates as clinical course, outcome, family history, and treatment response," the *DSM-III* authors wrote.[26]

The APA's hope and expectation was that further research would val-idate the diagnoses. The manual's reliability, Klerman told the APA members, would put researchers into "a position to test theories about etiology."[27] Moreover, thanks to ongoing advances in neuroscience, the APA and academic psychiatrists were optimistic that mental disorders would soon give up their biological secrets. New brain imaging tech-nologies, such as computerized axial tomography and positron emission tomography, were allowing scientists to "see" the brain in new ways, and researchers were also developing novel methods for studying how neurons in the brain communicated.

"I cannot imagine a time of greater excitement and potential in the field of psychiatry than the present," said APA president John Talbott in 1985. "After years of research in a variety of scientific and clinical areas, we seem to be on the verge of several major substantive breakthroughs."[28]

Indeed, the field was now hot on the trail of an exciting hypothesis: perhaps mental disorders were due to chemical imbalances in the brain. This hypothesis, which offered up the hope that the etiology of men-tal disorders was becoming known, had been born from discoveries about how psychiatric drugs act on the brain. Chlorpromazine and other antipsychotics had been found to block dopamine receptors in the brain, thereby hindering dopamine transmission; thus, researchers hypothesized that perhaps psychotic disorders were due to too much dopamine. The fact that amphetamines, which increased dopamine activity, could trigger psychosis lent support for this hypothesis. Meanwhile, the two classes of antidepressants, tricyclics and monoamine oxidase inhibitors, both inhib-ited the removal of serotonin and norepinephrine from the synaptic cleft (the tiny gap between neurons), which theoretically increased levels of these "monoamines" in the brain. This led researchers to hypothesize that perhaps depression was due to low levels of these brain chemicals.

After the publication of *DSM-III*, the public began to regularly hear of this possibility. In a 1981 article, which featured an interview with University of Chicago psychiatrist Herbert Meltzer, the Associated Press (AP) told readers that "researchers believe clinical depression is caused by a chemical imbalance in the brain."[29] Even more exciting, Meltzer told the AP that two new drugs were already in development that "restore the chemical imbalance in the brain." Other newspaper articles began talk-ing about a "revolution" that appeared to be unfolding in psychiatry, and then, in 1984, Nancy Andreasen, who would later become editor-in-chief of the *American Journal of Psychiatry*, captured the public's attention with a best-selling book, *The Broken Brain: The Biological Revolution in Psychiatry*. The new thinking in psychiatry, she wrote, was that "the major psychi-atric illnesses are diseases. They should be considered medical illnesses just as diabetes, heart disease and cancer are."[30] The thought was that each "different illness has a different specific cause," she said, adding that researchers were now honing in on those causes. "There are many hints that mental illness is due to chemical imbalances in the brain and that treatment involves correcting these chemical imbalances."[31]

That same year, *Baltimore Evening Sun* reporter Jon Franklin wrote a six-part series, titled "The Mind Fixers," which laid out this new frontier in psychiatry in even grander terms. "Psychiatry today stands on the threshold of becoming an exact science, as precise and quantifiable as molecular genetics," he wrote. Franklin interviewed more than 50 scientists and collectively they told of amazing breakthroughs that were around the corner. The field's understanding of the brain and mental illness, said Johns Hopkins psychiatrist Joseph Coyle, was moving ahead at such a dizzying pace that "we have gone from ignorance to almost a surfeit of knowledge in only ten years." The great discovery being made was this: "People who act crazy are acting that way because they have too much or too little of some chemicals that are in their brains," said NIMH scientist Candace Pert. "It's just physical illness!" Thanks to this new knowledge, psychiatry could now follow a simple strategy for developing new drugs: find the chemical basis of mental disease and devise ways to correct that. The day was drawing near, Franklin concluded, that "psychiatry would become capable of curing the mental diseases that afflict perhaps 20 percent of the population."[32]

Franklin was awarded a Pulitzer Prize in expository journalism for his series. It told of astonishing progress for a medical specialty that, only four years earlier, had acknowledged that the etiology of mental disorders was unknown. If this future came to pass, it would arguably make for an advance that surpassed any of medicine's grand triumphs of the twentieth century.

"Our field is exploding with new information, optimism, and enthusiasm," said APA president Carol Nadelson in 1985. "Psychiatry has moved from a backwater to the forefront as a medical specialty, largely because of the research explosion, particularly in the neurosciences."[33]

This understanding, that psychiatric drugs were being developed that fixed known chemical imbalances, became set in the public mind with the introduction of Prozac to the market in 1988. In so many ways, Prozac's arrival represented a realization of the revolution in care promised by Andreasen and Franklin, and as Eli Lilly marketed its new drug, the public was given a quick lesson in neuronal communication so they could appreciate why this was so.

A brain cell, known as a presynaptic neuron, releases serotonin into the minuscule gap between neurons (the synaptic cleft). This chemical messenger fits into receptors on a second neuron (the postsynaptic neuron) like a "key into a lock." This causes the second neuron to fire. The serotonin is then quickly removed from the synaptic cleft and returned to the presynaptic neuron via a "reuptake" process. Depression was caused by too little serotonin in the brain—presynaptic neurons released too little of it—and Prozac fixed this problem by inhibiting the reuptake of serotonin. Serotonin now stayed longer in the synaptic cleft than normal, and in this way upped serotonergic activity in the brain. As such, fluoxetine could be said to be "like insulin for diabetes," as it remedied a known deficiency. Even its name—a selective serotonin reuptake inhibitor (SSRI)—suggested a molecule that honed in on a precise pathology.

While Eli Lilly took the lead in telling this story to the public, both the NIMH and the APA helped push it along. Five months after Prozac hit the pharmacy shelves, the NIMH launched its Depression, Awareness, Recognition, and Treatment program (DART). The purpose of the campaign was to inform the public that depressive disorders "are common, serious, and treatable."[34] Eli Lilly helped pay for the printing and distribution of eight million DART brochures, which were distributed to physicians' offices around the country and, among other things, explained the particular merits of "serotonergic drugs" for treating depression.[35] The APA also mounted its own campaign about depression, and as the *St. Petersburg Times* told readers in August 1988, after interviewing the APA's John Talbott and Harvey Ruben, most patients would now be treated with drugs "that restore the chemical imbalance scientists have linked to many depressions."[36]

Prozac was the first of many SSRIs, and as other pharmaceutical companies brought their antidepressants to market, they all told this same story. Pfizer, GlaxoSmithKline, Forest Laboratories—their drugs were all said to remedy a chemical deficiency, and this was so even when they were prescribed for other mood disorders, like anxiety. "Chronic anxiety can be overwhelming. But it can also be overcome," a GlaxoSmithKline ad in *Newsweek* said. "Paxil, the most prescribed medication of its kind for generalized anxiety, works to correct the chemical imbalance believed to cause the disorder."[37]

In the mid-1990s, Janssen, Eli Lilly and other pharmaceutical companies brought new "atypical antipsychotics" to market (risperidone, olanzapine, etc.), and these drugs were also said to fix a chemical imbalance. The National Alliance for the Mentally Ill published a book titled *Breakthroughs in Antipsychotic Medications*, and in it psychiatrist Peter Weiden explained why these new drugs were an advance over the old ones: "Conventional antipsychotics all do about the same job in the brain. They all correct brain chemistry by working on the dopamine systems in the brain...the newer medications seem to do a better job of balancing all of the brain chemicals, including dopamine and serotonin."[38]

The APA, in its public information campaigns, focused on informing the public that depression and other diagnoses were "real illnesses" and that "treatment works." In that regard, it mostly stood aside and let both the pharmaceutical companies and the National Alliance for the Mentally Ill—and other advocacy groups that received funding from pharmaceutical companies—tell of drugs that fixed chemical imbalances in the brain. But, on occasion, it also directly endorsed this belief, and celebrated its spread in the American public. In a 2001 article that appeared in *Family Circle* magazine, which was part of a "special advertising feature," APA president Richard Harding wrote, "In the last decade, neuroscience and psychiatric research has begun to unlock the brain's secrets. We now know that mental illnesses—such as depression or schizophrenia—are not 'moral weaknesses' or 'imagined' but real diseases caused by abnormalities of brain structure and imbalances of chemicals in the brain."[39] In that same issue of *Family Circle,* Nada Stotland, a professor of psychiatry at Rush Medical College, who would subsequently become president of the APA, wrote that antidepressants "restore brain chemistry to normal."[40]

The APA, in this public forum, was clearly promoting the chemical imbalance story. Four years later, in a May 2005 press release, the APA noted that, in a recent survey it had conducted, "75 percent of consumers believe that mental illnesses are usually caused by a chemical imbalance in the brain." This, said APA president Steven Sharfstein, was evidence of "good news for [public] understanding of mental health." The problem, the APA stated, was that "those surveyed are almost twice as likely to seek help from a primary care physician rather than a psychiatrist," even though the psychiatrist was "a specialist specifically trained to diagnose and treat chemical imbalances."[41] That same year, the APA published its "Let's Talk Facts About Depression" brochure, which delivered the same message: "Antidepressants may be prescribed to correct imbalances in the levels of chemicals in the brain."[42]

At that point, the APA, pharmaceutical companies, and the National Alliance for the Mentally Ill had spent nearly two decades informing the public about the biology of mental disorders, and a 2006 survey documented the end result: 87 percent of Americans now knew that schizophrenia was caused by a chemical imbalance in the brain, and 80 percent understood that was true for major depression too.[43] This understanding fit into a larger story of scientific achievement. In 1980, the APA had published a diagnostic manual that had been proven to be "reliable," and in the ensuing years psychiatric researchers had validated that manual in the most convincing way possible: they had discovered the biological causes of major mental disorders. That was the story of a scientific triumph of the first rank.

The Search for Chemical Imbalances

In the scientific literature, however, a very different story about chemical imbalances had played out. As noted above, the theory was born in the 1960s, after researchers discovered how antipsychotics, monoamine oxidase inhibitors, and tricyclics acted on the brain. In order to test the hypothesis that these drugs were correcting a chemical imbalance, researchers needed to investigate whether depressed patients actually had low levels of monoamines (e.g., serotonin or norepinephrine) in the brain, or whether schizophrenia patients suffered from overactive dopamine systems.

Although researchers had no way to directly measure neurotransmitter levels in living patients, they seized upon a novel method to indirectly do so. In the 1960s, researchers had discovered that there were two ways that a neurotransmitter was removed from the synaptic cleft: either it was taken back up by the presynaptic neuron, or an enzyme would metabolize it and the metabolite would be removed as waste. Scientists found they could isolate the metabolite in the cerebrospinal fluid, and they reasoned that measuring metabolite levels could provide an indirect measure of neurotransmitter activity in the brain. Dopamine is metabolized into homovanillic acid (HVA), and thus if a person had too much dopamine activity, then the amount of HVA in his or her cerebrospinal fluid should be higher than normal. Serotonin is metabolized into 5-hydroxyindole acetic acid (5-HIAA); thus, if a person suffered from too little serotonin, then the amount of 5-HIAA in his or her cerebrospinal fluid should be lower than normal.[44]

Researchers first measured 5-HIAA levels in depressed patients in the late 1960s and early 1970s, and right from that first moment, the monoamine theory of depression began to fall apart. In 1971, investigators at McGill University reported that they had failed to find a "statistically significant" difference in the 5-HIAA levels of depressed patients and normal controls.[45] Three years later, Malcolm Bowers at Yale University reported the same finding. Serotonergic levels in the brain seemed normal, at least by this measure.[46] Researchers also tried to investigate the theory by giving nondepressed people monoamine-depleting drugs, reasoning that if low levels of monoamines caused depression, this should induce depression. But when two investigators at the University of Pennsylvania, Joseph Mendels and Alan Frazer, reviewed the scientific literature, they found that this wasn't the case. The subjects in the experiments had not reliably become depressed. "The literature reviewed here strongly suggests that the depletion of brain norepinephrine, dopamine, or serotonin is in itself not sufficient to account for the development of the clinical syndrome of depression," they wrote in 1974.[47]

In 1984, NIMH studied another possibility: The 5-HIAA levels of depressed patients did fall along a bell curve (as was the case for "normals"), and so perhaps those at the low end of the curve constituted a biological subgroup, who could be said to suffer from low serotonin, and thus it should be this group that would respond the best to an antidepressant, amitriptyline, that selectively blocked the reuptake of serotonin (and thus increased serotonin levels in the synaptic cleft). However, the researchers found that those with high 5-HIAA levels were just as likely to respond to amitriptyline as those with low levels. The NIMH drew the obvious conclusion: "Elevations or decrements in the functioning of serotonergic systems per se are not likely to be associated with depression."[48]

Even after that 1984 report, investigators continued to investigate whether depressed patients suffered from low serotonin, with this research quickening after Prozac arrived on the market in 1988. However, the studies, time and again, failed to find evidence that it was so. The third edition of the APA's *Textbook of Psychiatry,* which was published in 1999, traced this research history, and pointed out the faulty logic that had led to the chemical imbalance theory of depression in the first place:

> The monoamine hypothesis, which was first proposed in 1965, holds that monoamines such as norepinephrine and 5-HT [serotonin] are deficient in depression and that the action of antidepressants depends on increasing the synaptic availability of these monoamines. The monoamine hypothesis was based on observations that that antidepressants block reuptake inhibition of norepinephrine, 5-HT, and/or dopamine. However, inferring neurotransmitter pathophysiology from an observed action of a class of medications on neurotransmitter availability is similar to concluding that because aspirin causes gastrointestinal bleeding, headaches are caused by too much blood and the therapeutic action of aspirin in headaches involves blood loss. Additional experience has not confirmed the monoamine depletion hypothesis.[49]

In short, the hypothesis that depression was due to low serotonin or to a deficiency in other monoamines had been investigated and found to be wanting. The passage in the 1999 textbook was the APA's acknowledgement of this fact, and in the ensuing years, a number of experts in the field made the same point. In his 2000 textbook *Essential Psychopharmacology,* psychiatrist Stephen Stahl wrote that "there is no clear and convincing evidence that monoamine deficiency accounts for depression; that is, there is no 'real' monoamine deficit."[50] Finally, Eric Nestler, a scientist famous for his investigations into the biology of mental disorders, detailed in a 2010 paper how the many types of investigation into this theory had all come to the same conclusion:

> After more than a decade of PET studies (positioned aptly to quantitatively measure receptor and transporter numbers and occupancy), monoamine depletion studies (which transiently and experimentally reduce brain monoamine levels), as well as genetic association analyses examining polymorphisms in monoaminergic genes, there is little evidence to implicate true deficits in serotonergic, noradrenergic, or dopaminergic neurotransmission in the pathophysiology of depression. This is not surprising, as there is no *a priori* reason that the mechanism of action of a treatment is the opposite of disease pathophysiology.[51]

For nearly 40 years, science had been telling a consistent story, and yet it was at odds with what the APA—and psychiatry as a medical specialty—had been leading the public to believe. In a 2012 program on National Public Radio (NPR), host Alex Spiegel observed that the idea that "depression is caused by a chemical imbalance in the brain" remained "popular," and it was then, in that mass media venue, that the public heard the truth. "Chemical imbalance is sort of last-century thinking," explained Joseph Coyle, editor-in-chief of *Archives of General Psychiatry*. "It's much more complicated than that…It's really an outmoded way of thinking." Alan Frazer, chair of the pharmacology department at the University of Texas Health Science Center, told NPR listeners what must have seemed like a startling fact: "I don't think there's any convincing body of data that anybody has ever found that depression is associated, to a significant extent, with loss of serotonin."[52]

No convincing evidence *ever* found, and yet this is precisely what the American public knew to be true.

While scientific investigations into the dopamine hypothesis of schizophrenia produced a more nuanced story, the hypothesis that antipsychotics fix a known chemical imbalance, and thus could be likened to insulin for diabetes, was largely seen as a discredited, or at least overly simplistic, theory by the early 1990s. First, in the 1970s, Malcolm Bowers and others measured levels of dopamine metabolites in the cerebral spinal fluid of schizophrenia patients, and found that, prior to exposure to an antipsychotic, their metabolite levels "were not significantly different from controls."[53] At that point, investigators turned their attention to a second possibility. Perhaps people diagnosed with schizophrenia had too many dopaminergic receptors, and that is what made their brains "hypersensitive to dopamine." In 1978, Philip Seeman at the University of Toronto reported in *Nature* that he had found that to be true. At autopsy, the brains of 20 schizophrenics had 70 percent more D_2 receptors than normal (the D_2 receptor is one of many subtypes of dopamine receptors, and is the subtype most strongly blocked by antipsychotics). However, all of the patients had been on antipsychotics, and Seeman confessed that this abnormality might "have resulted from the long-term administration of neuroleptics."[54] Subsequent investigations found that to be the case, with investigators in France, Sweden, and Finland all reporting that there were no "significant differences" in D_2 receptor densities in living patients who had never been exposed to neuroleptics and "normal controls."[55]

There was an obvious irony in this finding. Researchers had hypothesized that schizophrenia was due to too many dopamine receptors in the brain, and yet they had discovered that while patients did not regularly suffer from this abnormality before exposure to neuroleptics, they often did after being treated with the medications. The drugs induced the very abnormality hypothesized to cause psychosis. In the early 1980s, researchers put together an understanding of why this occurred: antipsychotics block D_2 receptors, and in an effort to compensate for that blockade, the brain increases the density of those receptors. The brain is trying to maintain the normal functioning of its dopaminergic pathways.

In the wake of these findings, a number of researchers concluded that that dopamine hypothesis, at least in its simplest form, had not panned out. There is "no good evidence for the perturbation of the dopamine function in schizophrenia," noted John Kane, a well-known psychiatrist at Long Island Jewish Medical Center, in 1994.[56] Seven years later, Eric Nestler and former NIMH director Steve Hyman, in their book, *Molecular Neuropharmacology,* reiterated this point: "There is no compelling evidence that a lesion in the dopamine system is a primary cause of schizophrenia."[57]

However, a number of researchers have continued investigating dopamine function in schizophrenia patients, reasoning that perhaps they suffered from dopamine abnormalities in particular regions of the brain. One thought is that people diagnosed with schizophrenia have too much dopamine in the brain stem area and too little in the frontal lobes. However, those investigations do not support the notion that schizophrenia is due to a hyperactive dopamine system in all brain regions, which is then brought into balance by antipsychotics. In 2012, two Swedish investigators, Aurelija Jucaite and Svante Nyberg, summarized the latest thinking in the field:

> Vigorous search for abnormalities in the dopamine system in schizophrenia so far has yielded inconclusive results. The increasing understanding of the behavioral complexity of schizophrenia suggests that it is unlikely that a single neurotransmitter system can explain such diverse symptoms, for example, inattention and hallucinations. Thus, any simple, exclusive pathology of the dopamine system was and is doubtful.[58]

The low-serotonin theory of depression and the dopamine hyperactivity theory of schizophrenia provided the foundation for a more universal "chemical imbalance" theory of mental disorders, and when these two disease-specific cases did not support this theory, most researchers began to think it was unlikely to be true for other disorders. The limited investigations that were done in that regard, such as studies of attention deficit hyperactivity disorder (ADHD), also failed to find evidence to support this theory. In 2005, Kenneth Kendler, coeditor-in-chief of *Psychological Medicine,* provided a succinct epitaph for this long search: "We have hunted for big simple neurochemical explanations for psychiatric disorders and not found them."[59]

Since then, many prominent psychiatrists have made similar confessions. "Earlier notions of mental disorders as chemical imbalances," wrote NIMH director Thomas Insel in a 2011 blog, "are beginning to look antiquated."[60] Insel said that it now appeared that mental disorders were "disorders of brain circuits," and under his leadership, psychiatry is now moving on to new theories about the biological causes of psychiatric disorders. But as that occurred, it produced an awkward moment: Why had the public been led to think, for so long, that psychiatric drugs fixed chemical imbalances in the brain, when science hadn't shown it to be so?

In 2011, Ronald Pies, editor of *Psychiatric Times,* sought to answer that question:

> I am not one who easily loses his temper, but I confess to experiencing markedly increased limbic activity whenever I hear someone proclaim, "Psychiatrists think all mental disorders are due to a chemical imbalance!" In the past 30 years, I don't believe I have ever heard a knowledgeable, well-trained psychiatrist make such a preposterous claim, except perhaps to mock it. On the other hand, the "chemical imbalance" trope has been tossed around a great deal by opponents of psychiatry, who mendaciously attribute the phrase to psychiatrists themselves. And, yes—the "chemical imbalance" image has been vigorously promoted by some pharmaceutical companies, often to the detriment of our patients' understanding. In truth, the "chemical imbalance" notion was always a kind of urban legend— never a theory seriously propounded by well-informed psychiatrists.[61]

Psychiatry's one fault, Pies added in a later blog, was that "to be sure, those of us in academia should have done more to correct these [false] beliefs."[62]

Neither Reliable Nor Valid

When the APA published *DSM-III,* the thought was that future research would "validate" the disorders. The chemical imbalance story informed the public that psychiatric disorders had been "validated" in the most convincing way possible: the etiology of major mental disorders had been discovered, and psychiatric drugs provided a remedy to that known pathology. Although that didn't turn out to be the case, such that the etiology of major mental disorders remains unknown, it is still possible that the disorders in *DSM-III* and *DSM-IV* might have been validated, at least to some degree, according to the criteria Guze and Robins had set forth in the 1970s. And if that were so, this would be a significant scientific accomplishment, and it would mean that the APA's manual provided society with a scientifically meaningful way to think about mental disorders.

By the early 1990s, leaders in American psychiatry suggested that ongoing research was achieving that goal. In a 1991 article in *Minnesota Medicine*, NIMH director Lewis Judd declared that DSM had proven to be both "reliable and valid," such that psychiatric diagnoses could now be made with the same degree of certainty as diagnoses for such conditions as "diabetes mellitus."[63] *DSM-IV* was published in 1994, and the following year, APA president Mary England said that as a result, the "validity of diagnosis has been enhanced."[64] The editors of the *American Journal of Psychiatry* saw it the same way, writing in 1995 that although laboratory markers for psychiatric disorders had not emerged, the diagnoses nevertheless had been "validated by clinical description and epidemiological data...The validation of psychiatric diagnoses establishes them as real entities."[65] In 2007, APA president Carolyn Robinowitz was even more emphatic: "Mental disorders are [now] recognized as real illnesses," she said, adding that just as "cancer is highly treatable and can be cured, we are experiencing a similar success in psychiatry."[66]

The APA, in its educational campaigns, conveyed this same message to the public. Research had shown that psychiatric disorders were "real illnesses." They weren't simply constructs, which involved grouping people with similar symptoms together for research purposes, but validated diseases.

However, as the APA's *DSM-5* task force began work in 2002, psychiatry had reason to look closely at this question of validity, and it became evident to most, including members of the task force, that two decades of research had failed to provide data that met the Robins–Guze validating criteria. The etiology of mental disorders remained unknown. The field still did not have a biological marker or a genetic test that could be used for diagnostic purposes. Psychiatric patients were "co-morbid" for many diagnoses, which belied the Robins–Guze criteria that validated disorders should be easily distinguished from one another. Drug treatments did not produce disease-specific responses, but rather produced similar responses in people with different diagnoses. In a 2012 roundtable discussion of the DSM, which involved more than 20 experts in psychiatric diagnostics, these facts were reviewed, which led the moderator, Yale Medical School psychiatrist James Phillips, to conclude that "virtually all discussants" agreed that "most of the diagnoses fail the test of the original Robins and Guze...validators."[67]

Phillips then raised this profound question: "The startling failure of research to validate the DSM categories of *DSM-III* and *DSM-IV* has led to a conceptual crisis in our nosology: what exactly is the status of DSM diagnoses? Do they identify real diseases, or are they merely convenient (and at times arbitrary) ways of grouping psychiatric symptoms?"[68]

That is a question that psychiatry is still struggling to answer. But as leaders in the field have confronted it, many have spoken of how the disorders have not been validated, which, in fact, provides a reason to question the DSM's utility.

> The strength of each of the editions of DSM has been "reliability"— each edition has ensured that clinicians use the same term in the same ways. The weakness is its lack of validity. —NIMH director Thomas Insel, 2013.[69]

> DSM diagnoses are not useful for research because of their lack of validity...DSM diagnoses have given researchers a common nomenclature—but probably the wrong one. —Nancy Andreasen, former editor of the *American Journal of Psychiatry,* 2007.[70]

[DSM disorders] no longer seem at all reducible to simple diseases, but rather are better understood as no more than currently convenient constructs or heuristics that allow us to communicate with one another as we conduct our clinical, research, educational, forensic, and administrative work. —Allen Frances, chair of the DSM IV task force, 2012.[71]

In clinical settings it is widely acknowledged that the phenotypes with which actual patients present bear only mild resemblance to those that define individual DSM entities...The current approach to psychiatric nosology is simply dying of its own flaws. —G. Scott Waterman, University of Vermont Department of Psychiatry, 2012.[72]

While in principle, science should be the basis of any diagnostic system, DSM has been seduced by the illusion that advances in neuroscience provide empirical validity for a new system. In reality, we do not know whether conditions like schizophrenia, bipolar disorder, or obsessive compulsive disorder are true diseases. —Joel Paris, McGill University Department of Psychiatry, 2012.[73]

When I graduated a generation ago, I accepted *DSM IV* as if it were the truth. I trusted that my elders would put the truth first, and then compromise for practical purposes where they had no truths to follow. It took me two decades to realize a painful truth, spoken now frankly by those who gave us *DSM III* when Ronald Reagan was elected, and by those who gave us *DSM IV* when Bill Clinton was president: the leaders of those DSMs don't believe there are scientific truths in psychiatric diagnosis—only mutually agreed upon falsehoods. They call it reliability. —Nassir Ghaemi, Tufts Medical School Department of Psychiatry, 2013.[74]

These words all tell the tale of a failed "revolution." The American public may now understand that psychiatric disorders are "real illnesses," and that the DSM is a scientific guide to distinguishing such illnesses, but, as can be clearly seen, that is an understanding that arose from APA publicity efforts, and not from a faithful accounting of scientific findings. Most of the diagnoses in the APA's manual have yet to be "validated."

An Ethical Failure

It is easy to see that guild influences shaped the story that the APA told to the public about the reliability and validity of its diagnostic manual. It was a story that benefited the psychiatric profession in many ways, and in particular, helped legitimize psychiatry as a medical specialty in the public mind. The problem is that the APA, in telling that story, failed in its duty as a *medical specialty* to faithfully communicate to the public what science was revealing about the reliability and validity of its DSM manual, and that the APA helped promote a false notion—that psychiatric drugs corrected chemical imbalances in the brain—which became fixed in the public mind.

In sum, this is a history that shows a medical profession putting its guild interests ahead of its duty to society and to its patients.

CHAPTER FIVE

Psychiatry's New Drugs

As the saying goes, "if one tortures the data long enough, it will eventually confess to anything."

—Erick Turner, 2013[1]

In the two decades following the publication of *DSM-III*, American psychiatry, in concert with pharmaceutical companies, brought a second generation of psychiatric drugs to market, which were heralded as much better than the older drugs. A new benzodiazepine, Xanax, was approved for the treatment of panic disorder, which had been newly identified as a discrete disorder in *DSM-III*. Prozac and other SSRIs took the nation by storm and quickly replaced tricyclics and monoamine oxidase inhibitors as the antidepressants of choice. Next, atypical antipsychotics like Risperdal and Zyprexa were touted as breakthrough medications, both more effective and much safer than chlorpromazine, haloperidol, and other "standard neuroleptics." Our society embraced the use of the second-generation drugs, such that by 2010, one in every five Americans was taking a psychotropic medication on a daily basis.[2]

Our focus, in this chapter, is on the role that psychiatry played in the process of testing these drugs and informing the public of their merits. If we turn the clock back to the 1980s, it is easy to define our societal expectation at that time: Our expectation was that researchers would adhere to the principles of good science when conducting the trials, and that, following Food and Drug Administration (FDA) approval, academic psychiatrists and the American Psychiatric Association (APA) would accurately detail those results in their communications to the public. In that way, both the academic community and the APA would fulfill their obligation to serve the public good.

The scientific standards that are supposed to govern the testing of an experimental drug are well known. Double-blind randomized control trials are the gold standard for assessing the efficacy of medications. The trial should be designed to provide a fair comparison of the experimental drug versus placebo, or the experimental drug versus an older drug. The primary outcome measures used to assess efficacy should be identified before the trial is started. The statistical methods that will be employed to analyze the data should, if possible, be identified a priori. If secondary outcomes are to be measured, they too should be identified prior to the start of the study. At times, researchers may conduct a post hoc analysis in order to identify an unexpected result, perhaps in different subgroups of patients, which can legitimately advance scientific knowledge. However, if such analysis is done with the purpose of identifying a result that favors the study drug, it is known as data mining, which, as many have noted, is akin to throwing a dart at a wall and then drawing a bulls-eye around the dart afterward.

The authors of a published study are asserting intellectual ownership of the paper. If they simply sign off on a paper written by the employees of a pharmaceutical firm (or by a medical writing company hired by the firm), they are participating in a ghostwriting exercise. The results, of course, should be presented in an accurate, thorough manner. Finally, as the investigators speak to the public about the merits of the experimental drug, their comments should accurately reflect the trial data.

As can be seen, this entire process—from drug testing to public pronouncements about the drug—is supposed to be marked by scientific integrity at every step. If an independent researcher is subsequently provided access to the study protocol and raw data, that person should be able to document a chain of information—from raw data to scientific paper to public pronouncement—that tells a consistent, scientific story.

With this ethical context in mind, we can now investigate whether academic psychiatry and the APA, as they collaborated with pharmaceutical companies in the development of new psychotropic drugs, consistently adhered to these scientific standards. This is a critical function of a medical specialty, for it is a standard that must be met if good science is to guide our societal use of medical treatments.

Xanax: A New Drug for Panic Disorder

In 1981, the FDA approved alprazolam, which Upjohn sold as Xanax, as a treatment for anxiety. Alprazolam is a benzodiazepine, and as a treatment for anxiety, it was a "me too" drug. Benzodiazepines had been used to treat anxiety for two decades, with Valium the most widely prescribed medication in the Western world from 1968 to 1981.[3] However, in the mid-1970s, the alarm had been sounded over their addictive properties, with Senator Edward Kennedy, in a hearing on the dangers of benzodiazepines, declaring that these drugs had "produced a nightmare of dependence and addiction, both very difficult to treat and recover from."[4] The number of prescriptions for benzodiazepines dropped from 103 million in 1975 to 71 million in 1980, with authorities in both the United States and the United Kingdom urging that their use be limited to the short term.[5]

Upjohn marketed Xanax as having "safer" qualities than Valium. It was a short-acting tranquilizer, with a half-life of only 6 to 12 hours, whereas Valium's half-life was 20 to 100 hours, and thus, Upjohn said, this new drug "tranquilized" users for a shorter period of time.[6] The implication was that it might also be less addictive than Valium, but there was little scientific evidence that this was so. Still, its arrival did help stabilize the market for benzodiazepines. Then Upjohn, shortly after it obtained FDA approval to sell Xanax as an antianxiety agent, identified a potential new market for its drug.

In *DSM-III*, the APA had categorized panic disorder as a discrete illness for the first time. This was now a disease without any FDA-approved drug treatment, and Upjohn saw an opportunity to make Xanax the first. Upjohn hired Gerald Klerman, a former director of the NIMH, to codirect the steering committees that would direct the clinical trials of alprazolam for panic disorder.

This was a project that Klerman and the APA saw as an important opportunity for psychiatry. The testing of Xanax would help legitimatize panic disorder as a real illness, and by extension, help validate the APA's new diagnostic manual. While Upjohn would fund the study, Klerman and other academic investigators would put their intellectual stamp on it. The steering committees that Klerman cochaired regularly met to review the protocol and make amendments, and to develop plans for analyzing the data and the "scientific presentation" of the results. "These governance groups provided a set of mechanisms to promote the effective collaboration between the pharmaceutical company and the academic community," Klerman said.[7]

The Cross-National Collaborative Panic Study: Design and Results

The Cross-National Collaborative Panic Study, which compared alprazolam to placebo, had two parts. First, efficacy was assessed over eight weeks of treatment, with this trial conducted at eight sites. Then, at two of the eight sites, there was a withdrawal study: both the alprazolam and placebo groups were tapered from their respective pills, over a period of four weeks, and then followed for an additional two weeks. This withdrawal component reflected the fact that a benzodiazepine, because of its addictive properties, was supposed to be used only on a short-term basis, and thus any treatment regimen needed to include weaning from the drug. The expectation was that the alprazolam patients would be doing better than the placebo group at the end of eight weeks and also at the end of 14 weeks, as this would show the drug was effective for curbing panic symptoms, and that patients could be safely withdrawn.

Benzodiazepines are known to work fast, and that proved to be the case in this study. The number of panic attacks that the alprazolam patients suffered decreased dramatically in the first week, and by the end of four weeks, the alprazolam patients were doing significantly better, in terms of a reduction in symptoms, than the placebo patients. However, during the next four weeks, the placebo patients—at least those who remained in the study—continued to get better, with their improvement during this period more pronounced than improvement in the alprazolam patients, such that by the end of eight weeks, there were "no significant differences between the groups" on most of the rating scales. This included total panic attacks per week, percentage of time panic free, and various quality-of-life measures.[8]

As might be expected, given that a benzodiazepine dampens the activity of the central nervous system, adverse events were common in the alprazolam group. These patients frequently suffered sedation (50 percent), fatigue, slurred speech, amnesia, and poor coordination.[9]

During the tapering phase, the alprazolam patients worsened dramatically, while the placebo patients continued to improve. Thirty-nine percent of the alprazolam-withdrawn patients "deteriorated significantly" and had to be put back on the drug. Thirty-five percent suffered rebound "panic" or "anxiety" symptoms more severe than they had experienced at baseline. An equal percentage suffered a host of debilitating new symptoms, including confusion, heightened sensory perceptions, depression, a feeling that insects were crawling over them, muscle cramps, blurred vision, diarrhea, decreased appetite, and weight loss. By the end of the six-week discontinuation study, the alprazolam patients were suffering, on average, 6.8 panic attacks per week, while the placebo patients, who had continued to improve during this period, were experiencing 1.8 attacks each week.[10]

These findings led to an evident conclusion. Once the outcomes from the tapering phase were considered, the Cross-National Collaborative Panic Study told of a drug that, on the whole, did more harm than good.

The Story in the Scientific Literature

Even before the study results were published, Klerman noted, "the initial findings of the efficacy of alprazolam for panic disorder" had already been "widely disseminated at various meetings and symposia." Those initial reports informed the medical community of a drug that was a "safe and effective" treatment for panic disorder, and then in the May 1988 issue of *Archives of General Psychiatry,* the "detailed findings" were presented.[11] Upjohn had hired *Archives* editor Daniel X. Freedman as a consultant, and now the journal devoted much of its May issue to Xanax reports.[12] Gerald Klerman penned an introductory article, while the academic investigators who had led the trials published three separate reports: one on short-term efficacy, one on the drug's safety, and one on the discontinuation arm of the study. The authors reported faculty affiliations with the University of Iowa, University of California at Los Angeles (UCLA), University of South Carolina, Mount Sinai School of Medicine in New York, and at academic institutions in Canada and Australia. As Klerman emphasized in his overview, it was these academic psychiatrists, as opposed to Upjohn, who were responsible for the published findings.

"The investigators participated in the statistical analysis and took responsibility for professional presentations and scientific publications," Klerman wrote. The papers published in the *Archives* "express the joint conclusions of the principal investigators and other members of the steering committee and have the imprimatur of the investigators."[13]

Although the short-term efficacy study was designed to assess outcomes at the end of eight weeks, the investigators, both in the abstract of their published article and in the discussion section, focused instead on the four-week results, which, in the abstract, they now referred to as the "primary comparison point." At that midway point in the trial, 82 percent of the alprazolam patients were rated "moderately improved or better," versus 43 percent of the placebo group. As noted above, at eight weeks there was no statistically significant difference in panic symptoms and most other outcomes between the alprazolam patients and the placebo patients who completed the study, but the investigators did not include that result in the abstract. In essence, the investigators had switched endpoints for the study, and by doing so had turned a negative result into one that told of alprazolam's robust efficacy.

The investigators then turned their attention to the safety data. Although the alprazolam patients experienced a number of adverse events, with 24 percent experiencing "impaired mentation" at the end of eight weeks, the researchers concluded that "on the whole, subjects were relatively free of side effects, and a substantial proportion reported none at all."[14]

Such were the shortcomings of the two published reports on alprazolam's short-term efficacy and safety. But the real challenge for the investigators, if they were to maintain an evidence-based rationale for prescribing the drug, was to present the results from the discontinuation phase in a way that didn't tell of harm done. The data told of drug-withdrawn patients who experienced worse anxiety and panic than at baseline; of patients who had ended up addicted to Xanax; and of patients who were suffering four times as many panic attacks as the placebo group by the end. While those results could be found in the published data, in the abstract of their article, the researchers reported a different finding: At the end of the discontinuation study, they wrote, the "outcome scores" for the alprazolam-withdrawn patients "were not significantly different from those of the placebo-treated group"[15] (Figures 5.1 and 5.2).

To draw that conclusion, the researchers had isolated the scores of a subset of patients in the alprazolam-withdrawn group: those who had managed to complete the study. In this way, they left out the scores for the 23 alprazolam-withdrawn patients who had deteriorated so severely they had to be put back on the drug. Then, when the researchers compared the scores of the 61 percent of the drug-withdrawn patients who had completed the study to the placebo group on several secondary outcomes scales—those that measured anxiety, phobia, and the physician's global assessment of the patients—they found that while the placebo group did in fact have better scores on all three measures, because of the small number of patients left

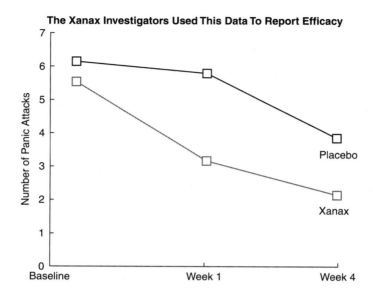

Figure 5.1   The study of Xanax was designed to measure the efficacy of the drug versus placebo at the end of eight weeks. However, the investigators emphasized the results at the end of four weeks, when the drug provided a statistically significant benefit.

Source: C. Ballenger, "Alprazolam in panic disorder and agoraphobia," *Arch Gen Psychiatry Psychiatry* 45 (1988): 413–22.

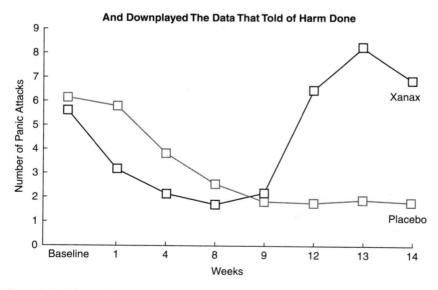

Figure 5.2   After eight weeks, those treated with Xanax were withdrawn from the medication. As this chart shows, the Xanax patients were much worse than the placebo patients at the end of the study.

Source: C. Ballenger, "Alprazolam in panic disorder and agoraphobia," *Arch Gen Psychiatry* 45 (1988):413–22. Also, C. Pecknold. "Alprazolam in panic disorder and agoraphobia." *Arch Gen Psychiatry* 45 (1988): 429–36.

in the study, the differences weren't statistically significant. However, the researchers had still left out the most critical finding: the difference in the number of panic attacks during the final week of the study between the two groups. This was the *target* symptom to be measured, and that difference was still statistically significant. That result was not mentioned in the abstract or discussed in the text; instead, it had to be extracted from a table.

Selling Panic Disorder

Once the study results were published, which, Klerman wrote, provided a "demonstration of the efficacy of alprazolam compared with placebo in the short-term treatment of panic disorder," the APA and its leaders helped Upjohn build a market for Xanax.[16] To do so, they sought to make doctors and the general public more aware of this new disorder and of Xanax's effectiveness.

The APA first reached out to its own members. In 1987, Robert Pasnau, who had been head of the APA in 1987, sent a glossy booklet on the *Consequences of Anxiety* to APA members, an "educational" effort paid for by Upjohn. The following year, Upjohn sponsored a symposium at the APA's annual meeting, where the "expert panel" highlighted the four-week results from the Cross National Collaborative Panic Study. Next, after the FDA approved Xanax as a treatment for panic disorder in 1990, Shervert Frazier and Gerald Klerman penned a "Dear Doctor" letter that Upjohn included in the promotional literature it sent to doctors, including general practitioners, about prescribing Xanax for this purpose.[17] Finally, the NIMH sponsored a conference on panic disorder, where a panel of experts designated "high potency benzodiazepines"—this would be Xanax—as one of the two "treatments of choice."[18]

The APA utilized a variety of media to reach the public. As we noted in chapter 3, the APA—with funding from Upjohn—produced two films, *Panic Prison* and *Faces of Anxiety,* with the first film premiering in 1989, just before the FDA approved Xanax for panic attacks. The APA also used an Upjohn grant to develop a "workshop on panic for nonpsychiatric professionals," and to conduct a general educational campaign, with the APA distributing "mental illness awareness guides" to educators and "nonpsychiatric health professionals" that told of how anxiety and panic disorder were "under-recognized and undertreated."[19] In a letter to the *New York Times,* the APA described these efforts as "a responsible ethical partnership that uses the no-strings resources of one partner [Upjohn] and the expertise of the other to help the one person in five who needs help and hope in struggling with mental illness."[20]

Thanks to this partnership, when the FDA approved of Xanax as a treatment for panic disorder, the public learned of a remarkably effective new drug. "In a Panic? Help Is On the Way," a *St. Louis Post Dispatch* headline announced.[21] Xanax, the paper reported, works for "70 percent to 90 percent" of those who suffer from the illness. Four million

Americans were said to suffer from the disorder, and early in 1992, the *Union Leader* newspaper in Manchester, New Hampshire reported that it "wasn't until recently that panic disorders became widely recognized," with clinics now opening that specialized in treating it.[22]

All of this made Xanax a remarkable commercial success. In 1993, it became the fifth most frequently prescribed drug in the United States.

Prozac, The Wonder Drug

The story of the selling of Prozac, starting with the checkered history of the clinical trials of Prozac, has been told in detail by a number of authors. However, those accounts focus on Eli Lilly's behavior and explain the drug company's misdeeds. What we want to do is examine these trials to see whether there is evidence of the institution of academic psychiatry becoming corrupted by the influence of pharmaceutical funding (and guild interests). We also want to look at the actions of the APA through this same lens. Is there evidence of those twin economies of influence affecting its educational campaigns about depression, and in ways that might be seen as "corrupt?" If so, then the trials of Prozac can be seen as an early moment in the development of a new normative behavior within psychiatry.

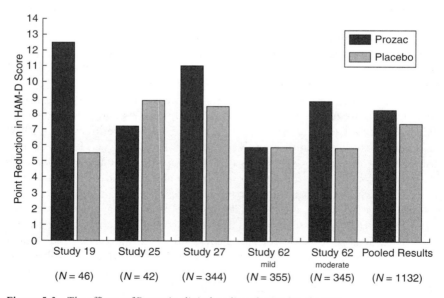

Figure 5.3   The efficacy of Prozac in clinical studies submitted to the FDA.

While academic psychiatry helped promote Prozac as a "breakthrough" antidepressant, the drug barely beat placebo in clinical trials. As the pooled results from the five studies show, Prozac reduced symptoms on the HAM-D scale only one point more than placebo, which is a clinically meaningless difference.

Source: I. Kirsch. "The emperor's new drugs: an analysis of antidepressant medication data submitted to the U.S. Food and Drug Administration." *Prevention & Treatment* 5, July 15, 2002.

As part of its new drug application for fluoxetine, Eli Lilly presented data from five clinical studies to the FDA (Figure 5.3). Reviewers of the protocols for these studies could immediately see that there were several design flaws that seriously undermined the value of the results. The biggest problem was that Eli Lilly had allowed investigators to prescribe benzodiazepines to control agitation that was frequently found in the Prozac-exposed group, and there were also patients in the study who were taking a sedative to help them sleep at night. This concurrent use of other psychiatric medications, admitted Eli Lilly's Dorothy Dobbs in a later legal case, was "scientifically bad," since it would "confound the results," and "interfere with the analysis of both safety and efficacy."[23] In addition, noted FDA reviewer David Graham, Eli Lilly had engaged in "large-scale underreporting" of the harm that fluoxetine could cause.[24] Even with the concurrent use of other psychiatric medications, Eli Lilly struggled to present data showing that its drug was efficacious. Two of the five clinical studies did not show a statistically significant benefit for the drug, and for a third to be seen as positive, the data had to be parsed in a certain way.[25] One of the remaining studies, known as protocol 27, compared fluoxetine to imipramine; the FDA concluded that the data showed that "imipramine was clearly more effective than placebo, whereas fluoxetine was less consistently better than placebo."[26] Indeed, when the outcomes from all five studies were pooled, the improvement in HAM-D scores for the fluoxetine patients was only one point more than for the placebo patients, which, by any measure, was a clinically meaningless difference.[27]

The Published Results

In his role as an expert witness in civil lawsuits that were eventually filed against Eli Lilly, psychiatrist Peter Breggin had an opportunity to see the FDA's reviews of Eli Lilly's five studies, which he then detailed in his book *Talking Back to Prozac*. In particular, Breggin identified the six principal investigators in protocol 27 (one of the five studies), and the FDA's analysis of their results. Four of the six investigators published their results in medical journals, which provides an opportunity to compare the FDA reviews with what they published, and determine if they are consistent.

- In 1984, psychiatrist James Bremner, who was the clinical director of Northwest Psychopharmacology Research, in Olympia, Washington, reported that at his site, "fluoxetine treatment significantly improved all measurements over baseline and was significantly more effective than imipramine in treating the depression and anxiety of the study patients."[28] In contrast, the FDA had determined that at his site, while fluoxetine was better than placebo on most variables, imipramine appeared more effective than fluoxetine, showing improvement on all variables.[29]

- In 1985, UCLA psychiatrist Jay Cohen reported that "fluoxetine relieved the symptoms of major depressive illness effectively and significantly better than placebo and was better tolerated than imipramine."[30] The FDA saw the data differently, noting that at his site, the blind had been broken at two weeks, instead of at six weeks, which led to a comparison of a response to fluoxetine at six weeks to response to placebo or imipramine at two weeks. This was an unfair comparison, and given this fact, an FDA reviewer concluded, "this study can, at best, be said to be supportive" of fluoxetine's effectiveness.[31]
- In a 1988 article, Utah psychiatrist Bernard Grosser reported that at his site, the "antidepressant efficacy of fluoxetine was superior to placebo and appeared to be similar to that of imipramine."[32] Once again, the FDA viewed the results differently, concluding that at Grosser's site, imipramine had produced "significantly more improvement than placebo on all major efficacy measures at endpoint," whereas fluoxetine had not been consistently better than placebo.[33]
- The final investigator to publish his results from protocol 27 was John Feighner, head of the Feighner Research Institute in California. In the abstract of his 1989 article, he reported that his results supported "previous studies which suggest fluoxetine's superior side effect profile and the approximate antidepressant equivalence of fluoxetine and TCAS (tricyclics)."[34] However, the FDA determined that at his site, imipramine was consistently better than placebo in "all variables," whereas "fluoxetine was not shown to be consistently different than placebo."[35]

This site-by-site comparision reveals that the published literature told a story at odds with the data, at least as reviewed by the FDA. The FDA, in its reviews of the results from all six sites in protocol 27, determined that imipramine had proven to be more consistently effective than placebo, and that this six-site protocol was "supportive but not strongly positive in demonstrating fluoxetine's role in the treatment of depression."[36] In contrast, the reports published by the four investigators in that study told of how fluoxetine was regularly superior to placebo, and equal in efficacy to imipramine, and that fluoxetine had a better safety profile than imipramine. In that way, they told of a therapeutic advance, even though the data didn't support that finding.

Their reports became the "scientific" finding regularly repeated by other psychiatrists in their reviews of this new drug. Fluoxetine was equal in efficacy to imipramine and more tolerable, and soon several psychiatrists added a new "fact" to this story: "Studies have shown that [serotonin] deficiency plays an important role in the psychobiology of depressive illness," wrote Sidney Levine, in a 1987 article in the *British Journal of Psychiatry*.[37] Fluoxetine, since it theoretically raised serotonin levels, might be fixing a chemical imbalance. Indeed, wrote two University of Louisville psychiatrists, since depression appeared to be due to a serotonergic deficiency, fluoxetine was "an ideal drug for the treatment of depression."[38]

Selling Depression

In 1985, Shervert Frazier, who had previously served as the chairman of the APA's Joint Commission on Public Affairs, was appointed director of the NIMH. In his role at the APA, he had helped lead the organization's efforts to improve its public image, and to promote its *DSM-III* disease model. Once he moved into his NIMH office, he announced plans for a "major educational campaign" designed to "change public attitudes" about depression.[39] The problem, from the NIMH's point of view, was that few Americans sought treatment when they became depressed. A 1986 poll quantified this concern: 78 percent of adults said they would not seek any treatment, explaining that they would simply live with the depression "until it passed."[40] Only 12 percent said they would take an antidepressant.

The NIMH launched its Depression Awareness Recognition and Treatment campaign (DART) in the spring of 1988, just as Prozac was brought to market. The campaign's primary message was that depressive disorders were "unrecognized, untreated, and undertreated," and that it was vital that people seek out medical help. "Left untreated," the NIMH now informed the public, "depression may be a fatal disease." Fortunately, antidepressants were highly effective. "Recovery rates with these medications have been shown to be in the range of 70 percent to 80 percent in comparison with 20 percent to 40 percent for placebo," the public learned.★[41]

The NIMH turned DART into a joint "private and public effort." Pharmaceutical companies, the APA and other organizations were all invited to contribute "resources, knowledge, and other forms of assistance to the project." DART developed educational materials for the public, primary care physicians, and mental health professionals, all of which were designed to increase their recognition of depressive disorders, and thus get more people into treatment.[42] With pharmaceutical support, the APA sponsored training sessions for primary care physicians and other mental health providers, where a particular emphasis was placed on "biological and pharmacological treatments."[43] Eli Lilly, with its new drug on

---

★ This efficacy claim would have surprised Jonathan Cole, the former director of the NIMH's Psychopharmacology Service Center. In a 1969 review, he found that antidepressants provided only a marginal benefit over placebo: In well-controlled studies 61% of medicated patients had improved, versus 46% of placebo patients, leading Cole to conclude that the "differences between the effectiveness of antidepressant drugs and placebo are not impressive." In addition, in 1986, the NIMH had run its own study of antidepressants, which compared imipramine to two forms of psychotherapy and to placebo and had found that at the end of 16 weeks, "there were no significant differences among treatments, including placebo plus clinical management, for the less severely depressed and functionally impaired patients." Only the severely depressed patients had fared better on imipramine than on placebo.

the market, distributed eight million pamphlets titled "Depression: What You Need to Know" through the DART program, with this "public service" brochure informing the public, among other things, of the particular merits of serotonergic drugs.[44]

Together, the reports that appeared in the medical literature, the DART campaign, and Eli Lilly's marketing of Prozac provided a foundation for newspaper and television stories that told of a new breakthrough drug for depression, a disease that struck so many more Americans than had ever been realized. This was a disease that for too long had gone unrecognized and undertreated, and a new SSRI antidepressant had arrived that was so much better than imipramine and the older tricyclics. *Newsweek* hailed it as a "Wonder Drug for Depression," while *The New York Times* described it as "one of the best antidepressants ever designed."[45,46] *Sixty Minutes* told of how most doctors now believed that chronic depression was "caused by a chemical imbalance in the brain," which could be corrected by the new drug. "Today depression can be treated—quickly and effectively—in seven cases out of ten," *Time* reported in 1992. "If a second round of treatment is required, the cure rate jumps to 90 percent."[47]

The data, of course, revealed a much different story about Prozac. Academic psychiatry and the APA could have noted, in their public comments, that the clinical trials were flawed in their design, and that even with those flaws, the trials showed that fluoxetine provided at best a very slight benefit over placebo. This new drug appeared to be less effective than the old standby, imipramine. Yet, the medical community learned of a new drug that was as effective as imipramine and much safer, and the public was informed of a breakthrough medication that fixed a chemical imbalance in the brain, and thus could "cure" 70 percent or more of all patients.

Zoloft: Discordance between Data and Published Results Becomes the Norm

One of the elements of "institutional corruption" is that problematic behaviors become normalized within the institution and thus accepted. In our reviews of Xanax and Prozac, we saw how, in articles published in medical journals, the results were presented in ways that overstated their benefits and glossed over their adverse effects. After Prozac, the next SSRI to be approved for sale in the United States was Pfizer's sertraline, sold as Zoloft, and once again, the reports in the scientific literature provided a misleading picture of its merits.

As had been the case with fluoxetine, clinical tests of sertraline did not provide robust evidence of its efficacy. In four of the six trials submitted by Pfizer to the FDA, sertraline failed to produce a better result than placebo. Both trials in hospitalized patients failed. There was a fifth study that was "questionable," and a sixth that was positive for sertraline. The collective results were so iffy that Pfizer struggled to convince European regulatory bodies to approve sertraline. As a Pfizer employee confessed in an April 11, 1991 memorandum, sertraline had "received an unfavorable review in a number of countries. The common key issue is that regulators are not convinced of sertraline's efficacy versus placebo."[48]

Since a drug manufacturer is supposed to provide two studies that show efficacy, the FDA's Psychopharmacological Drugs Advisory Committee, which met on November 19, 1990, focused in its review on the merits of Pfizer's "questionable" one. That study compared three fixed doses of sertraline—50 mg, 100 mg, and 200 mg—to placebo, and at the end of four weeks, there was no difference in symptom reduction at any of the doses compared to placebo. However, during the last two weeks, the 50 mg group improved more than placebo, such that there was a statistically significant difference compared to placebo at six weeks in this one group. That was the extent of the evidence for efficacy in the questionable study, a two-week period of greater efficacy than placebo in one of the three sertraline groups, and even then, the difference between drug and placebo was only 2.3 points on the HAM-D scale. This difference, noted psychiatrist Jeffrey Lieberman, from Albert Einstein College of Medicine, "is not necessarily robust enough to be clinically effective."[49]

The advisory committee struggled to come to a decision. Even if the questionable study was deemed positive, said Robert Hamer, a professor of psychiatry at the Medical College of Virginia, "how do we interpret those two positive results in the context of several more studies that fail to demonstrate the effect?"[50] The FDA's Paul Leber, while urging the committee to vote to approve sertraline, confessed that he had "no idea what constitutes proof of efficacy, except on the basis of what we, as a Committee, agree on, on an ad hoc [basis], that there needs to be."[51] The advisory committee, he added, "might say to us, 'look, we think the standards in this field are terrible. People have been getting away with non-substantive efficacy for years.'"[52]

In that room, experts in clinical trials were discussing the obvious: these trials did not provide strong evidence that sertraline was effective as a treatment for depression. Perhaps the two studies—one positive and one questionable—met a lax FDA standard for approval, but that was the best that could be argued. At last, the advisory committee voted six to three to approve sertraline. As for the questionable study, which the FDA ultimately accepted as a second "positive" study, an FDA reviewer explicitly noted its limitations, writing that "no unequivocal conclusion can be made about the relative efficacy of different Zoloft doses."[53] Only the 50 mg dose of sertraline, and not the 100 mg and 200 mg doses, had proven to be superior in the final two weeks of that study.

The FDA's understanding of sertraline's questionable efficacy did not make it into the published literature. The four negative studies weren't published, and the questionable study was cast as a positive one. The authors of that latter study, which included psychiatrists from Wayne

State University, University of Minnesota, and Vanderbilt University, reported that "for the evaluable-patient analysis, all sertraline groups showed significantly greater improvements in all efficacy variables except one when compared with the placebo group." *All* doses of sertraline had proven effective—this was precisely the finding that the FDA, in its analysis, said couldn't be made.

Here is how the authors were able to make this claim. The telling words are "evaluable-patient analysis." In the abstract, the authors were reporting outcomes only for patients who "were still taking study medication on or after day 11 of the double-blind treatment phases." This knocked out a lot of poor responders to sertraline, and thanks to this data-mining analysis, the authors could report that "all dosages of sertraline were substantially more effective than placebo in treating major depression."[54]

That was one of many published studies that told of sertraline's efficacy for depression. One such follow-up report was authored by Oregon Health Sciences University psychiatrist Daniel Casey, who had chaired the FDA's advisory committee for sertraline, when its efficacy had been much questioned. "The existing scientific evidence indicates that sertraline is effective for managing moderate and severe depression both in acute and maintenance phase. Sertraline has a wide therapeutic index," he wrote.[55] In a later disclosure statement, Casey reported that he was a consultant to Pfizer and on its speaker's bureau.[56]

A "Dirty Little Secret" Revealed

As other pharmaceutical companies brought SSRIs to market, the same basic scenario was repeated. Many of the trials failed to show a benefit for the drug, but the negative studies either went unpublished or were spun into positive findings. Celexa and Paxil were the next two SSRIs that were approved by the FDA, and at that point, with four SSRIs on the market, the FDA had assessed the merits of 31 studies for these four drugs, and even with the FDA's charitable standards for determining that a study was "positive"—it would often allow the company to sort through the data, in a post-hoc analysis, to find a positive outcome—only 14 were positive. There were 14 others that were negative, and three more that were questionable. However, the published literature related to those 31 studies told of 19 positive outcomes and two negative ones.[57] The medical literature simply didn't reflect, in any meaningful way, what clinical trials had revealed about the efficacy of the four drugs, and this was just the tip of the iceberg. Ghostwritten papers from post-marketing studies also filled psychiatric journals, with these reports regularly telling of the drugs' efficacy, and this ghostwriting practice became so accepted that SmithKline Beecham, as it marketed Paxil, organized a campaign called "Case-Study Publications for Peer Review," which it wittily dubbed CASPPER, mindful of television's friendly ghost.[58]

There were, of course, many academic researchers who knew of the discrepancy between the raw data submitted to the FDA and the published literature. Psychologist Irving Kirsch, after he obtained the FDA's reviews of the SSRI trials, immediately recognized this discrepancy. "The small difference between the drug response and the placebo response has been a 'dirty little secret' known to researchers who conduct clinical trials, FDA reviewers, and a small group of critics who analyzed the published data and reached conclusions similar to ours," he noted in 2002. "It was not known to the general public, depressed patients, or even their physicians."[59]

This "dirty little secret" did finally come out, thanks to Kirsch and others. In 2008, Erick Turner from Oregon Health and Sciences University, in a review of FDA data for 12 antidepressants approved between 1987 and 2004, reported that 36 of the 74 studies had failed to find any statistical benefit for the antidepressants, and that of these 36 negative or questionable studies, 22 went unpublished, 11 were spun into positive results, and only three were properly reported.[60] That same year, Kirsch reported that in the trials of Prozac, Effexor, Serzone, and Paxil, symptoms in the medicated patients dropped 9.6 points on the Hamilton Rating Scale of Depression, versus 7.8 points for the placebo group, a difference of only 1.8 points. The National Institute of Clinical Excellence in Britain had previously determined that a three-point difference on the Hamilton scale was needed to demonstrate a clinically significant difference. Kirsch found that it was only for the severely depressed patients, with a baseline Hamilton score above 25, that the SSRIs had provided a benefit that met this 3-point standard.[61]

With this information having finally come out, the *British Journal of Psychiatry* confessed that clinical trials had, in fact, generated "limited valid evidence" supporting the prescribing of antidepressants, except in severe cases.[62] Newspapers and magazines treated these reports as stunning news—it was really quite baffling to understand how this could be—and *Sixty Minutes*, which 20 years earlier had told of Prozac's wonders, now ran an "expose" on this astonishing fact.[63] All along, the data emerging from clinical trials had actually provided reason to question the merits of the SSRIs, but that understanding, for the longest time, was absent from the medical literature.

The Atypicals: More of the Same

Within the conceptual framework of institutional corruption, the pharmaceutical industry could be said to have "captured" academic psychiatry and the APA as it tested and marketed the SSRI antidepressants. The pharmaceutical companies hired psychiatrists from prestigious medical schools to serve as their key opinion leaders, and industry also provided funds to the APA as it ran educational campaigns that informed the public that depression often went unrecognized and untreated, and that there were highly effective drug treatments for this illness. As pharmaceutical companies brought new antipsychotics to market, the corrupting effects of this industry "capture" of psychiatry can be seen once again.

The Risperidone Files

The notion of an "atypical" antipsychotic was born when Sandoz brought clozapine to market, selling it as Clozaril. This drug blocked both serotonin and dopamine receptors, and didn't cause the usual high incidence of extrapyramidal symptoms (EPS) that haloperidol and the first-generation antipsychotics did. However, because it could cause agranulocytosis, a potentially fatal depletion of white blood cells, the FDA approved it in 1989 only as a second-line therapy for patients who didn't respond to standard antipsychotics. In 1994, the FDA approved Janssen's new antipsychotic, risperidone, and this marked the moment that the era of "atypical" antipsychotics really took hold.

Janssen conducted three "well-controlled" trials to support its New Drug Application for risperidone.[64] In the first, involving 160 patients at eight US centers, risperidone was superior to placebo in reducing positive symptoms. However, it wasn't better than placebo on the "Clinical Global Impression Scale," which measures overall improvement. Nearly 50 percent of the risperidone patients dropped out of the six-week study. Janssen's second study, which involved 523 patients at 26 sites in the United States and Canada, compared four doses of risperidone to a 20 mg dose of haloperidol and to placebo. Janssen argued that this study showed that its drug, at an optimal dose of 6 mg, was superior to haloperidol for treating positive and negative symptoms, but FDA reviewers noted that Janssen had used a single, high dose of haloperidol for comparison, a dose that "may have exceeded the therapeutic window" for some patients, and thus the study was "incapable by virtue of its design of supporting any externally valid conclusion about the relative performance of haloperidol and Risperdal."

The FDA determined that Janssen's third study, involving 1,557 patients in 15 foreign countries, was similarly flawed. Janssen compared five doses of risperidone to a 10 mg dose of haloperidol, with Janssen once again claiming that this showed its new drug to be more effective than the old, but Paul Leber, director of the FDA's Division of Neuropharmacologic Drugs, concluded that the study, by its very design, was "incapable" of making any meaningful comparison. In order to honestly compare two drugs, an equal number of "equieffective" doses must be tested, as otherwise the study unfairly favors the drug that is given in multiple doses. Such trial design, Leber wrote on December 21, 1993, is "a critical preliminary step to any valid comparison of their properties."

In sum, the FDA concluded that Janssen had shown evidence that risperidone was superior to placebo in reducing positive symptoms, but that it hadn't provided evidence that it was any better than haloperidol. In the FDA's letter of approval to Janssen, Robert Temple, director of the FDA's Office of Drug Evaluation, made this clear:

> We would consider any advertisement or promotion labeling for RISPERDAL false, misleading, or lacking fair balance under section 502 (a) and 502 (n) of the ACT if there is presentation of data that conveys the impression that risperidone is superior to haloperidol or any other marketed antipsychotic drug product with regard to safety or effectiveness.[65]

While that may have been the FDA's conclusion, readers of medical journals were coming to a different understanding of Risperdal's merits. Within a year of its approval, there were at least 30 articles in medical journals that reported on clinical tests of this new drug, and together they told of an important therapeutic advance for the treatment of schizophrenia. Risperidone was said to be equal or superior to haloperidol in reducing psychotic symptoms, and superior to haloperidol in improving negative symptoms (lack of emotion). Researchers reported that, in comparison to the older drugs, risperidone reduced hospital stays, improved patients' ability to function socially, and reduced hostility.[66] The most compelling article of all reported on the results from Janssen's 523-patient trial. This was authored by Steven Marder and John Davis, who were known as top schizophrenia experts, and many of the 26 investigators who conducted the study had academic credentials. The data analysis "was performed independently by Dr. Davis without support from the Janssen Research Foundation." This was an article that could be expected to have a big impact on the field, and here is what they concluded: "Risperidone produced significantly greater improvements than haloperidol on all five dimensions" of schizophrenia. It was more effective than haloperidol in reducing positive symptoms, and there were "large" differences between the two drugs on "negative symptoms, hostility/excitement, and anxiety/depression."[67]

Their article spoke of a new drug that was globally better than haloperidol, and it appeared in the literature four years after Robert Temple informed Janssen that the FDA would consider such claims, if made in an advertisement, to be "false and misleading." However, the FDA's review remained out of sight, and with such glowing reports of risperidone appearing in the scientific literature, newspapers and magazines published stories of a breakthrough medication. Risperdal, the *Washington Post* wrote, "represents a glimmer of hope for a disease that until recently had been considered hopeless." Meanwhile, the *New York Times* provided its readers with a scientific explanation for why it was so effective: Risperdal relieved "schizophrenia symptoms by blocking excessive flows of serotonin or dopamine, or both."[68]

Zyprexa

Eli Lilly was the next company to seek approval for an atypical antipsy-chotic, and the FDA's review of its trials struck a familiar chord.[69] Leber and a second FDA official, Paul Andreason, found that Eli Lilly's studies of olanzapine were "biased against haloperidol" in much the same way that Janssen's had been. Multiple doses of olanzapine had been compared to a single dose of haloperidol, and the drugs were not compared at "equief-fective" doses. The FDA also noted that Eli Lilly had not excluded patients who previously had responded badly to haloperidol, and by including such patients, it would make the haloperidol results worse than normal. "The sample of patients used is an inappropriate choice" for comparison pur-poses, Leber wrote. For this reason and others, the FDA concluded that Eli Lilly's two large studies, including its large phase III study, involving 1,996 patients, provided little useful efficacy data. The trial data, Leber wrote, "do not provide a useful quantitative evidence of how effective (even in the short run) olanzapine actually will be in the population for whom it is likely to be prescribed." The company had provided some evi-dence that it was better than placebo over the short run, and thus could be approved, but its trial data was "insufficient to permit [Eli Lilly] to make claims asserting the product's superiority to haloperidol."

The FDA's critique of Eli Lilly's trials did not appear in the research literature. Instead, once again the published articles told of a drug that was better than haloperidol and other first-generation antipsychotics.[70] Psychiatry had produced a second breakthrough medication. Then, in interviews with the media, well-known psychiatrists told of how olanzapine, marketed as Zyprexa, appeared to be even better than Risperdal. John Zajecka, at Rush Medical College in Chicago, told the *Wall Street Journal* it was a "wonderful drug for psychotic patients," while Harvard Medical School's William Glazer said that "the real world is finding that Zyprexa has fewer extrapyramidal side effects than Risperdal."[71] Stanford University psychiatrist Alan Schatzberg told the *New York Times* that Zyprexa was "a potential breakthrough of tremen-dous magnitude."[72]

Seroquel

The third atypical approved by the FDA was Astra Zeneca's quetiapine (Seroquel). This time, the FDA reviewers were, if anything, more critical of the quetiapine data than they had been of the risperidone and olanzap-ine results.[73] Four of the eight studies were not considered by the FDA to provide any "meaningful" efficacy data, and in the remaining four, while quetiapine was modestly superior to placebo for reducing psychotic symp-toms, it appeared less effective than haloperidol in this regard. Moreover, patients clearly had trouble staying on the new drug. Eighty percent of the 2,162 quetiapine-treated patients had dropped out of the trials, compared to 61 percent of the placebo patients and 42 percent of those treated with standard neuroleptics.

Given these results, AstraZeneca's authors had a more difficult time crafting "better than haloperidol" articles for submission to medical journals. Internal AstraZeneca reports concluded that haloperidol, risperidone, and other first-generation antipsychotics were more effective than quetiapine in reducing psychotic symptoms, and on several other outcome measurements.[74] Thus, at least initially, a more modest story was published: in 1997, the year that the FDA approved quetiapine, the Seroquel Trial 13 Study group, which was composed of psychiatrists from at least eight medical schools, published a paper in *Biological Psychiatry* that stated, in the abstract, that quetiapine was equal to haloperidol in efficacy, and well tolerated.

This pivotal study, involving 361 patients, compared five doses of quetiapine to one dose of haloperidol. Fifty-six percent of the quetiapine patients withdrew before the end of six weeks, and "mean duration of treatment" for the quetiapine patients was only 23 days, which was less than for either the placebo or haloperidol group. While quetiapine was comparable to haloperidol in knocking down psychotic symptoms, it was inferior to haloperidol, *at all five doses,* on improving negative symptoms. The data indicated that a majority of patients couldn't tolerate quetiapine for any length of time, with "lack of efficacy" being the primary reason that patients had withdrawn, and that the drug was inferior to haloperidol in improving the emotional withdrawal that could prove so problematic to schizophrenia patients.

Yet, in their abstract, the Seroquel 13 Study Group announced this finding: quetiapine was comparable to haloperidol in reducing positive symptoms and in reducing negative symptoms "at a dose of 300 mg/day." [75] Only attentive readers of that North American study, with an eye for charts, would have realized that the authors had cherry-picked the one dose of quetiapine (out of five) that was somewhat comparable to haloperidol in reducing negative symptoms in order to make that claim of equivalency (Figure 5.4). The authors also claimed in the abstract that quetiapine was "well tolerated," and once again, only attentive readers of the entire article would have seen that such a finding was belied by the fact that, on average, quetiapine patients had stayed in the study for little more than three weeks.

As AstraZeneca's thought leaders published such articles, many within the company—as internal e-mails reveal—understood that the published results didn't square with the actual trial data, at least not in a substantive way. In a December 6, 1999 e-mail, Astra Zeneca's John Tumas summed up the ethical quandary:

> There has been a precedent set regarding "cherry picking" of data...That does not mean that we should continue to advocate this practice. There is growing pressure from outside the industry to provide access to all data resulting from clinical trials conducted by industry. Thus far, we have buried Trials 15, 31, 56, and are now

Quetiapine v. Haloperidol on Negative Symptoms

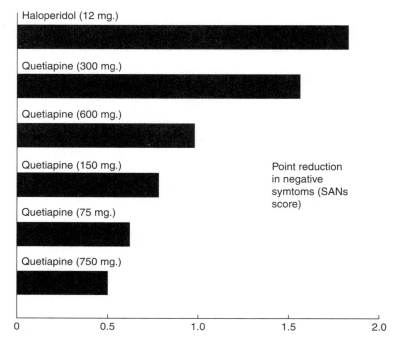

Figure 5.4   The finding in the published abstract was that quetiapine was comparable to haloper-idol "in reducing negative symptoms at a dose of 300 mg. per day." But as can be seen in this graph, quetiapine at all doses was less effective than haloperidol in reducing negative symptoms. However, the difference between haloperidol and quetiapine at 300 mg. was not statistically significant, which allowed the researchers to claim that quetiapine was "comparable" to haloperidol by focusing in the abstract on this one comparison, and omitting mention of the other four quetiapine doses.

Source: L. Arvanitis. "Multiple fixed doses of 'Seroquel' (quetiapine) in patients with acute exacerbation of schizophrenia: A comparison with haloperidol and placebo." *Biol Psychiatry* 42 (1997):233-246.

considering [burying] COSTAR. The larger issue is how do we face the outside world when they begin to criticize us for suppressing data. One could say our competitors indulge in this practice. However, until now, I believe we have been looked upon by the outside world favorably with regard to ethical behavior. We must decide if we wish to continue to enjoy this distinction.[76]

Three months later, Tumas had a specific case to worry about. The company had done a pooled analysis of four studies comparing quetiapine to haloperidol. "The data don't look good," he wrote in an internal e-mail. "I don't know how we can get a paper out of this." However, a short while later, at the 2000 annual meeting of the APA, Charles Schulz, a prominent psychiatrist at the University of Minnesota, who served as a key opinion leader for a number of pharmaceutical companies, including

AstraZeneca, announced this result: quetiapine had *bested* haloperidol in this pooled analysis of the four clinical trials. "I hope that our findings help physicians better understand the dramatic benefits of newer medications like Seroquel," Schulz said, "because, if they do, we may be able to ensure patients receive these medications first."[77]

Seroquel was the third atypical brought to market, and both the medical community and the public came to this general understanding: Risperdal, Zyprexa, and Seroquel were all atypical antipsychotics, a class of second-generation psychiatric drugs that were revolutionizing the treatment of schizophrenia. "It used to be that schizophrenics were given no hope of improving," the *Los Angeles Times* told its readers. "But now, thanks to new drugs and commitment, they're moving back into society like never before."[78] Meanwhile, the APA became directly involved in touting the merits of the new drugs, producing a "patient education video" with funds from Zeneca Pharmaceuticals.[79]

Dueling Narratives

In a sense, we can see three competing narratives in this review of the atypicals. The FDA did not find that the clinical trials of the three drugs provided evidence of a therapeutic advance over haloperidol. However, the medical journals told of two new drugs that were markedly superior to haloperidol, and another new drug that was equivalent in efficacy to haloperidol and better tolerated. Meanwhile, the reports in the media told of breakthrough medications. There was a vast disparity between the three narratives, and it took some time for the narratives to be reconciled, and it was the latter two, of course, that were shown to be false.

The first study in the medical literature to suggest that something might be amiss with the "breakthrough medication" story came rather quickly. In 1996, two Canadian researchers, Patricia Huston and David Moher, from the University of Ottawa, reviewed the literature on risperidone in order to do a meta-analysis of its efficacy, only to discover, much to their surprise, that the results from a single trial had been reported multiple times and by a multitude of authors. By the end of 1995, they noted in a *Lancet* article, results from the North American 26-center study had been reported "in part, transparently and not so transparently in six different publications," and "the authorship was different for each."[80] This "redundant publication" made it seem that a large number of studies had found risperidone superior to haloperidol. This was not how science was supposed to proceed, the researchers said.

> The use of different authors for the identical study in different publications obscures duplication. It is an obvious violation of the definition of authorship according to the International Committee of Medical Journals Editors. There should be concern not only about

gratuitous authorship, but also about ghost authorship...what does it say about clinical trialists who take part in these practices? It suggests either a large dose of naiveté or a fair degree of cynicism. Are some trialists not aware that redundancy is frowned upon? Do participating physicians realize that merely contributing to a trial is not a criterion for authorship?...Ultimately, everybody is affected by these practices. To abuse the honor system central to medical publications with the aim of overstating one's case undermines the integrity of science. It also calls into question author and investigator integrity.

In an accompanying commentary, *Lancet* editor-in-chief Richard Horton summarized the bad science in this way: "Individual centers reported their results both together and separately, previous reports of the same data were not cited, authorship changed with each publication, and patients were excluded from one report with no clear reason. The effect of this extraordinary publication activity was a greatly enhanced visibility for risperidone at a critical time in the drug's history; a marketing success, if nothing else."[81]

Those reports blemished the "breakthrough medication" glow that initially surrounded Risperdal. The "better than the old drugs" glow then slowly disappeared too. In 2000, a team of English investigators, led by John Geddes at the University of Oxford, reviewed the published reports of 52 randomized trials of the first six atypical antipsychotics to be marketed, and determined that "there is no clear evidence that atypical antipsychotics are more effective or are better tolerated than conventional antipsychotics." The most common ruse that had been employed to make the new drugs look better, Geddes noted, was the use of "excessive doses of the comparator drug."[82] This was the beginning of the research community's focus on the how the trials had been biased by design. Next, in 2003, a study by US Department of Veterans Affairs, led by Robert Rosenheck, found that Zyprexa did "not demonstrate advantages" compared with haloperidol in "compliance, symptoms, extrapyramidal symptoms, or overall quality of life."[83] Two years later came the CATIE bombshell. In this NIMH study, patients were randomized either to one of four atypicals or to an older, low-potency antipsychotic, perphenazine, and there were no significant differences between the older drugs and three of the four newer ones. (Zyprexa was seen as slightly better than perphenazine.) Even more worrisome, neither the new drugs nor the old ones could really be said to work. Seventy-four percent of the 1,432 patients were unable to stay on the medications to the end of the 18-month trial, mostly because of their "inefficacy or intolerable side effects."[84] Finally, in 2007, British researchers reported that schizophrenia patients, if anything, had a better "quality of life" on the old drugs.[85]

At that point, the atypical story had officially collapsed, and so completely that various researchers weighed in on what to make of it all. The published studies from the premarketing trials, confessed

Jeffrey Lieberman, "did not necessarily portray a clear picture of their effectiveness."[86] Psychiatrist Peter Jones, who had led Britain's study of the new drugs, said it wasn't quite right to say that he and his colleagues had been "duped," but rather that they had been "beguiled" by the early reports and the marketing claims.[87] It all told of a rather remarkable delusion that had taken hold in the international psychiatric community, which prompted two British researchers, Peter Tyrer and Tim Kendall, to write in *Lancet* of the "spurious advance of antipsychotic drug therapy."[88]

> What was seen as an advance 20 years ago—when a new generation of antipsychotic drugs with additional benefits and fewer adverse effects was introduced—is now, and only now, seen as a chimera that has passed spectacularly before our eyes before disappearing and leaving puzzlement and many questions in its wake...As a group they are no more efficacious, do not improve specific symptoms, have no clearly different side-effect profiles than the first-generation antipsychotics, and are less cost effective. The spurious invention of the atypicals can now be regarded as invention only, cleverly manipulated by the drug industry for marketing purposes and only now being exposed. But how is it that for nearly two decades we have, as some have put it, been beguiled into thinking they were superior?

Today, we can easily answer that question. The beguilement was rooted in the published reports, often authored by academic psychiatrists, that told of new drugs that were better than the old, and did so without discussing the design flaws in the trials. The research reports told the story that the drug companies wanted to tell, and the academic psychiatrists who conducted those studies and authored the articles provided the story "of new and improved drugs" with an apparent scientific legitimacy.

Not a Bad Apple, But a Bad Barrel

At the opening of this chapter we described societal expectations: We want the clinical testing of new drugs to be a scientific enterprise, designed to truly assess their merits, and we want the trial results to be accurately communicated in the research literature and to the public. But in this review, we found that trials were not always designed to provide an objective picture of the drug's harms and benefits; secondary endpoints and outcome measures were employed to present industry-friendly results; adverse events were minimized; and conclusions were drawn in abstracts that didn't accurately reflect the study data. Thus, we see in this history evidence of *institutional* corruption: organized psychiatry was incentivized to partner with industry, and practices and norms developed that compromised the institution's public health mission.

CHAPTER SIX

Expanding the Market

> Like knowledge itself, diagnostic language confers power, but that
> power is not necessarily benign.
>
> —Jeanne Maracek, 1993[1]

Once pharmaceutical companies obtained FDA approval for their new
psychiatric medications, they naturally wanted to maximize the sale of
their products. This meant expanding the pool of potential patients, and
to do this, the companies needed the assistance of academic psychiatry and
the American Psychiatric Association (APA). From the drug companies'
perspective, the business model to be pursued was obvious: They could
provide the financial resources for this task, while academic psychiatry
and the APA provided the medical legitimacy. This legitimacy, industry
knew, could originate with the DSM: The APA could expand the pool of
potential patients by creating new diagnoses or by loosening the diagnostic criteria for existing diagnoses, and then industry could hire academic
psychiatrists to conduct studies of the drugs for these new patient populations. Then industry could hire those same psychiatrists, or others in academia, as "key opinion leaders" to speak at professional conferences about
the validity of the "illness" and the helpful drug treatment.

The APA and academic psychiatry, of course, had their own guild reasons to push this effort. Expanding diagnostic boundaries would expand
the psychiatric enterprise. It would increase the potential pool of psychiatric patients; it would increase the influence of psychiatry in American
society; and it would provide new opportunities for academic psychiatrists to be recognized as experts in particular diagnoses. Industry money
would flow into all corners of psychiatry as well—to the APA, to psychiatric journals, and to the academic psychiatrists paid to serve as the key
opinion leaders. In 2002, Alan Gelenberg, editor-in-chief of the *Journal of
Clinical Psychiatry*, acknowledged this guild interest. Psychiatry, he wrote,
should work with industry to "grow the market."[2]

From an institutional-corruption perspective, the question to be examined in this chapter is this: Did the twin economies of influence that
were present during this exercise—that is, the pharmaceutical and guild

interests—corrupt this process? If psychiatry conducted research in a manner consistent with good clinical practices that revealed that this diagnostic expansion would provide a medical benefit to the American people, then psychiatry could be seen as fulfilling its public health mission. Indeed, this was part of Gelenberg's point: growing the market would bring helpful treatment to a larger segment of the American population. But if the science that was conducted as part of this process was compromised (or absent), then this market expansion can be seen as another example of how psychiatry, as a medical institution, was corrupted by these twin influences.

As we wrote earlier in this book, our focus is on how economies of influence may create a "bad barrel," as opposed to focusing on how such monetary rewards may lead an individual astray. But even from an institutional-corruption perspective, it is important to document the industry money that flowed to individual psychiatrists involved in setting diagnostic boundaries in order to fully show the presence and strength of this economy of influence in psychiatry as a whole.

The DSM: A Tool That Defines the Market

As noted in chapter 2, when Robert Spitzer and his task force were creating *DSM-III*, they adopted a policy of "syndromal inclusiveness." They wanted the manual to provide a diagnosis for everyone who might come into their offices. The DSM, Spitzer said, "defines what is the reality. It's the thing that says, 'this is our professional responsibility, this is what we deal with.'"[3] In other areas of medicine, diagnoses may be bounded by nature, with biological markers helping to distinguish disease states, but in psychiatry they are fungible, with the profession, in essence, able to set them wherever they want.

When Spitzer and his collaborators put together *DSM-III*, they increased the number of diagnoses to 265, 83 more than had been in *DSM-II*. They did so, scholars have noted, to avoid giving people who sought treatment a "false-negative" diagnosis—that is, a diagnosis that didn't truly reflect their symptoms, but one that had to be given because there was no diagnosis available that accurately described their symptoms. At the same time, this increase in diagnostic categories had the consequence of expanding the boundaries of what was to be considered a mental illness.

This increase in diagnostic categories was accomplished in two ways. In a number of instances, Spitzer and his collaborators divided a single *DSM-II* diagnosis into several new diagnoses. For instance, anxiety neurosis in *DSM-II* was cleaved into two disorders in *DSM-III*, panic disorder and generalized anxiety disorder. The diagnosis of "inadequate personality" in *DSM-II* was separated into three disorders: narcissistic personality disorder, borderline personality disorder, and dependent personality disorder. In addition, Spitzer and his *DSM-III* colleagues constructed a number of new diagnoses. For instance, they created four new diagnoses related to poor impulse control: pathological gambling, kleptomania, pyromania, and isolated explosive disorder. The APA continued this splitting and creating-new-disorders process when it published *DSM-IIIR* in 1987, with the diagnostic count rising to 292. Seven years later, with the publication of *DSM-IV,* the count jumped higher still, although various researchers have come up with different tallies, ranging from 297 to 374.

The public, when questioning the growth of the psychiatric enterprise over the past 30 years, has often focused on this increase in the number of disorders in each subsequent edition of the DSM. The fact that 297 or more diagnoses were listed in *DSM-IV* led to an obvious question: Could there really be that many discrete mental disorders? But in terms of expanding the patient pool, the more important mechanism has been the steady loosening of criteria needed to make a diagnosis for such common disorders as ADHD, depression, and bipolar disorder. This criteria-loosening process was critical to building expanded markets for stimulants, SSRIs, and atypical antipsychotics.

At first glance, it might seem that the APA, having adopted a disease model in 1980, would have found it difficult to expand diagnostic criteria in this way. In *DSM-III,* the disorders were presented as discrete illnesses. Since a diagnosis was based on whether a designated number of symptoms were present, the manual appeared to draw a clear line between illness and no illness. A diagnosis of major depression required the presence of five of nine symptoms said to be characteristic of the disorder; a person with only four of the nine was determined to not suffer from it. However, even as the authors of the DSM drew this sharp line between disease and no disease, they nevertheless subsequently noted, in a section on the "limitations of the categorical approach" in *DSM-IV,* that there was "no assumption that each category of mental disorder is a completely discrete entity with absolute boundaries dividing it from other mental disorders or from no mental disorder."[4]

This was a diagnostic approach that allowed psychiatry to have its cake and eat it too: the disorders could be conceptualized as specific illnesses, which would be understood as "diseases" of the brain, even as its "no absolute boundary" conception allowed for the expansion of diagnostic boundaries ever outward, into the sphere of normal behaviors and normal emotional difficulties. The diagnoses were wrapped in the cloak of medical diseases, even as the "limitations of the categorical approach" statement in *DSM-IV* told of boundary lines being arbitrarily drawn.

Many of the authors of the DSM manuals, as they drew up these diagnostic boundaries, had financial ties to industry, serving on advisory boards and as consultants to the pharmaceutical companies that made the products that would be sold to these diagnostic groups. At least 56 percent of the *DSM-IV* panel members had a financial tie to a drug company, and this conflict of interest was particularly pronounced on panels for psychiatric conditions where a medication was recommended as a first-line therapy. One hundred percent of the *DSM-IV* panels for "mood disorders" and for "schizophrenia and other psychotic disorders" had a conflict. Eighty-one percent of the "anxiety disorders" panel did, and 62 percent of the panel for "disorders usually first diagnosed in infancy, childhood or adolescence."[5]

DSM-IV was published in 1994, and based on its diagnostic criteria, researchers subsequently determined that 26.2 percent of American adults suffered "from a diagnosable mental illness in a given year."[6] This percentage, when applied to the 2010 US census, translated to 62 million adults. As for youth, *DSM-IV* needed 86 pages to describe the many disorders that could afflict children and teenagers, and based on the criteria set out there, the Centers for Disease Control and Prevention concluded that 13 percent of youth 8 to 15 years old experience a bout of mental illness each year.[7] If that percentage is applied to all youth 6 to 17 years old, that translates to about seven million children and teenagers, creating a total of 69 million Americans who suffered from a diagnosable "mental illness" in 2010.

This number could be understood as the potential market for psychiatric services based on the diagnostic criteria in *DSM-IV.* Add in the fact that physicians may "overdiagnose" certain common conditions, and it is clear that the APA's diagnostic manual, with its fungible boundaries, serves both industry and guild interests well.★

The Rise of ADHD

The diagnostic roots of ADHD are usually said to date back to 1902, when Sir George Frederick Still, a British pediatrician, published a series of articles on 20 children who were of normal intelligence but "exhibited violent outbursts, wanton mischievousness, destructiveness, and a lack of response to punishment."[8] Although Still couldn't identify any obvious illness or trauma that could explain their bad behavior, he nevertheless reasoned that it arose from a biological problem, and thus concluded that the children suffered from "minimal brain dysfunction."

Over the next 50 years, a handful of others seconded the notion that hyperactivity was a marker for mild brain damage. In 1947, Alfred Strauss, director of a school for disturbed youth in Racine, Wisconsin, called his extremely hyperactive students "normal brain injured children."[9] The APA, in *DSM-I*, said such children suffered from an "organic brain

---

★ As a number of scholars have noted, the reported effects of a drug can also drive the construction of new diagnostic categories. For instance, if a medication of any type (psychiatric or nonpsychiatric) is seen as relieving a distressing symptom or behavior, one that is not already a diagnosis, then there is now a rationale for creating such a category, as there is a drug available to effectively treat that distress.

syndrome," a term that was changed to "hyperkinetic reaction of child-hood" in *DSM-II*.

In 1956, Ciba-Geigy brought Ritalin (methylphenidate) to market as a treatment for narcolepsy, and soon physicians at Johns Hopkins University School of Medicine, aware of earlier reports in the scientific literature that stimulants could "subdue" children and help them focus better on school-work, deemed this new drug useful for quieting children who suffered from a "brain damage syndrome."[10] While there was no great rush by psychiatrists to prescribe this drug in the 1960s, usage did begin to climb in the 1970s, such that by the end of the decade, 150,000 children were taking methylphenidate.

Spitzer and his colleagues renamed this problem "attention deficit disor-der" (ADD) in *DSM-III*. The cardinal symptoms were said to be inatten-tion, impulsivity, and hyperactivity, although a diagnosis could be made even if hyperactivity wasn't present. A child who "often fails to finish things he or she starts," or "often acts before thinking," was now a candi-date for this diagnosis, provided such "symptoms" were causing the child difficulties in school or some other area of life. The disorder was said to be common, occurring in "as many as 3 percent of prepubertal children."[11] The APA renamed the diagnosis attention deficit hyperactivity disorder when it published *DSM-IIIR*, and reconfigured the diagnostic boundaries in ways that made it slightly easier to make the diagnosis.

The number of youth diagnosed with ADHD rose steadily during the 1980s, reaching 600,000 by the end of that decade, and then, in 1994, the APA expanded the diagnostic boundaries again. *DSM-IV* stated that the disorder consisted of three subtypes: inattentive only, hyperactive-impulsive only, and those who had both types of symptoms. If a child had six of nine characteristic symptoms of either subtype, a diagnosis of ADHD could be made, provided that the onset of such symptoms occurred by age seven.[12] Field trials of *DSM-IV* found that the new criteria, compared to *DSM-III* criteria, could be expected to lead to a 23 percent increase in the number of children with ADHD.[13] The disorder was now said to affect "3 percent to 5 percent" of all American children.

At that point, research into ADHD increased, as it now apparently affected several million youth. In particular, Harvard Medical School psychiatrist Joseph Biederman, who had been a member of the *DSM-IV* work panel for pediatric disorders, turned into a publishing machine. He began publishing articles on ADHD and other pediatric disorders at the rate of one new article every two weeks, a pace he continued for the next 20 years. Collectively, his ADHD reports told a comprehensive story.[14] ADHD was a real disorder, with an apparent genetic component, and there were indeed distinct subtypes, just as *DSM-IV* had indicated.[15,16] It was a chronic condition, with symptoms likely to persist into adulthood, and if ADHD youth were not treated, they were at high risk for multiple bad outcomes: poor academic achievement, failure in the workplace, substance abuse, criminal behavior, and mood disorders.[17,18] Stimulants lessened

ADHD symptoms, and they also improved "self-esteem, cognition, and social and family function," and lowered the risk of substance abuse.[19] In 1995, Biederman announced that 6 to 9 percent of American schoolchildren had ADHD, twice the prevalence rate listed in *DSM-IV*.[20]

This publication record turned Biederman into a well-known ADHD expert, who was often quoted by the media. In 1996, he told the *New York Times* that as many as 10 percent of the nation's children could benefit from Ritalin. Without such treatment, Biederman said, "all the follow-ups show bad outcomes [for ADHD youth], in any dimension you want."[21]

The marketing of ADHD soon received another boost, this time from the pharmaceutical industry. Ritalin had long been the treatment of choice, but it was an old drug, off patent, and thus there was no strong flow of industry dollars promoting this product. Then, in 1996, Shire obtained FDA approval for Adderall as a treatment for ADHD. Other new ADHD drugs were brought to market, and with these drugs under patent protection, their manufacturers now had millions of advertising dollars to spend. They sponsored symposiums at APA conferences; they filled the *Journal of the American Academy of Child and Adolescent Psychiatry* (and other psychiatric journals) with their advertisements; and they paid Biederman and other child psychiatrists to serve as their thought leaders.[22] From 1996 to 2011, Biederman received speaker's fees, consulting fees, and research funding from more than 24 pharmaceutical companies. His roster of clients included Shire, Janssen, and Eli Lilly, which respectively sold Adderall, Concerta, and Strattera, three of the most popular ADHD drugs.[23]

Such is the story of how the ADHD market was built. It wasn't a creation of the drug industry, but rather of organized psychiatry. *DSM-III* and *DSM-IV* provided the diagnostic framework, and academic psychiatrists published research that told of its validity and of the efficacy of ADHD medications. Pharmaceutical companies then popularized that scientific story, and it all led to a steady rise in ADHD diagnoses, such that in 2012, 10 percent of youth 4 to 18 years old had been so diagnosed. That year, 3.5 million American youth were prescribed an ADHD medication, nearly six times the number in 1990.[24]

As for whether this increased diagnosis and drug treatment provided a benefit to American youth, we'll delve into that question in the next chapter. The published findings by Biederman and a number of other researchers do tell of a medical advance, but a long-term study funded by the NIMH produced a different finding, as did other long-term studies. But these long-term findings, which so clearly threatened the ADHD market, aren't widely known by the America public, which is revealing in its own way.

Redefining Depression

In 1980, Americans filled about 30 million prescriptions for antidepressants.[25] Thirty-one years later, Americans filled 264 million prescriptions for this class of drugs, which was more than for any other class of medication.[26] The diagnostic changes that made this market expansion possible were of two types. In *DSM-III*, the APA defined major depression in a way that broke with long-standing medical tradition, which dramatically expanded who could be seen as clinically depressed, and it also carved anxiety into a number of disorders, each of which was said to be common. Together, these changes provided a framework for diagnosing millions of Americans who were experiencing emotional tumult of some sort, even though it could be seen as quite normal, with a disorder that could be treated with an SSRI.

In their book *The Loss of Sadness,* Allan Horwitz and Jerome Wakefield tell of the medical thinking that, beginning with the teachings of Hippocrates (460–337 BC), had long governed the diagnosis of depressive states.[27] Physicians, they noted, had always been careful to distinguish what they called melancholy from normal sadness. The former was a pathological condition, which arose without cause, or else was characterized by symptoms that were "out of proportion" to the circumstances and persisted for an abnormally long time. "Patients are dull or stern, dejected or unreasonably torpid, without any manifest cause; such is the commencement of melancholy," wrote the Greek physician Aretaeus of Cappadocia (ca. AD 150–200).[28]

Dejection that arose in response to a setback in life was understood by the ancient Greeks to be normal. Roman physicians made the same distinction, and so did the authors of medical texts in Renaissance Europe. For example, the English doctor Timothie Bright (1550–1615) wrote of the need to distinguish "betwixt melancholy and the soules proper anguish."[29] This understanding then crossed the Atlantic and was incorporated into the first American medical texts. Benjamin Rush (1745–1813), remembered as the father of American psychiatry, described melancholia as "directly contrary to truth, or it is disproportioned in its effects, or expected consequences, to the causes which induce them."[30]

Emil Kraepelin (1856–1912), the German psychiatrist whose diagnostic work is said to have inspired *DSM-III*, also wrote that in the absence of psychotic symptoms (which characterized most asylum patients), there was a need to distinguish between depressive illness and normal sadness. "Morbid emotions are distinguished from healthy emotions chiefly through the lack of a sufficient cause, as well as by their intensity and persistence," he wrote. "The severity, and more especially the duration, of the emotional depression [must] have gone beyond the limits of what is normal."[31]

Although Sigmund Freud offered psychiatry a different explanation for what caused a morbid depressive state, arguing that it had psychological origins rather than a biological cause, he too emphasized that depression needed to be distinguished from the grief and sadness that arose from losses in life. "Although grief involves grave departures from the normal attitude to life, it never occurs to us to regard it as a morbid condition and hand the mourner over to medical treatment. We rest assured that after a lapse of time it will be overcome, and we look upon any interference with it as inadvisable or even harmful."[32] The definition of depressive neurosis in *DSM-II* reflected this distinction: "This disorder is manifested by an excessive reaction of depression due to an internal conflict or to an identifiable event such as the loss of a love object or cherished possession."[33]

All of this long medical history reflected a common understanding: in order for a physician to make a diagnosis of clinical depression, the events of a person's life needed to be examined. Grief, sorrow, and anguish were emotions familiar to nearly everyone, as the works of poets, playwrights, and novelists could attest. However, in *DSM-III*, this age-old understanding was absent. The manual stated that if the person experienced a "dysphoric" mood, along with four other symptoms from a list of eight, for two weeks, then a diagnosis of depression could be made. There was no need for an inquiry into life events, except in the instance of "bereavement." If someone had suffered the loss of a loved one, then a diagnosis was not to be made, unless one's sorrow "was unduly severe or prolonged."[34]

As a result of this new diagnostic approach, a person who was going through a divorce, or who had lost a job, or was struggling to cope with any difficult life event—and was experiencing such "symptoms" as insomnia, fatigue, or poor appetite—could now be diagnosed with clinical depression. The authors of *DSM-III*, Horwitz and Wakefield wrote, "rejected the previous 2,500 years of clinical diagnostic tradition that explored the context and meaning of symptoms in deciding whether someone is suffering from intense normal sadness or a depressive disorder."[35]

This led to a dramatic increase in the percentage of Americans who could now be said to be "depressed." Prior to *DSM-III*, major depression was understood to be a fairly uncommon disorder. In her 1968 book *The Epidemiology of Depression,* Charlotte Silverman, who directed epidemiology studies for the NIMH, noted that a study of Maryland patients in 1963 produced a one-day prevalence rate, "for all depressive disorders," of less than one in 2,000 (for adults). The principal diagnostic categories, which were drawn from *DSM-I,* were "manic-depressive reaction" and "psychoneurotic depressive reaction." A study in 1966, which sought to ferret out depression seen in patients treated by general practitioners, found a one-year prevalence rate of about 1.2 percent.[36] After *DSM-III* was published, the NIMH's Epidemiologic Catchment Study, which was conducted in the 1980s, found that 5 percent of adults now experienced a depressive episode each year.[37] *DSM-IIIR* was published in 1987, and using the criteria in that manual, the NIMH's National Comorbidity Survey reported that 10 percent of the adult population suffered from major depression each year.[38] The prevalence of "major depression" had leapt tenfold—or more—in 20 years.

Those were the two epidemiological surveys that served as the scientific "evidence" for the APA's oft-repeated message that depression regularly went "undiagnosed and undertreated." If 10 percent of the American population suffered from major depression, and only 5 percent were in treatment, then there were millions of Americans who needed to get help. That was the message of the NIMH's DART program and numerous subsequent educational campaigns mounted by the APA, the NIMH, and other advocacy groups. In 1996, the National Depressive and Manic-Depressive Association, with funding from Bristol-Myers Squibb, convened a conference on this problem, and at its conclusion, a 20-member panel, which included many of the leading experts on mood disorders in academic psychiatry, authored a "Consensus Statement on the Undertreatment of Depression." The panel warned of a national health emergency:

> Depression is one of the most frequent of all medical illnesses. Depression is a pernicious illness associated with episodes of long duration, high rates of chronicity, relapse, and recurrence, psychosocial and physical impairment, and mortality and morbidity... There is overwhelming evidence that individuals with depression are being seriously undertreated. Safe, effective and economical treatments are available. The cost to individuals and society of this undertreatment is substantial.[39]

By the time that the National Depressive and Manic-Depressive Association sounded that alarm, in 1996, there was a diagnosis in the DSM for virtually anyone who complained of feeling a bit blue. "Dysthymia" was the diagnosis to be made when chronic depressive symptoms of a lesser intensity were present. "Adjustment Disorder with Depressed Mood" was for someone who exhibited "marked distress that is in excess of what would be expected from exposure to the stressor." Most inclusive of all, there was "Mood Disorder Not Otherwise Specified," which was for "disorders with mood symptoms that do not meet the criteria for any specific mood disorder."[40]

In this way, the APA developed a diagnostic framework that had the effect of expanding the market for antidepressants. The APA stopped distinguishing between event-related sadness and the melancholy that Hippocrates, Renaissance doctors, and Benjamin Rush had written about, and in *DSM-IIIR* and *DSM-IV*, it constructed categories of "lesser depression" that further expanded the pool of potential patients. However, as this particular market expansion occurred, there was little evidence presented that the SSRIs provided a benefit over placebo for those experiencing a mild to moderate bout of depression. In the absence of such data, this can only be seen as a development driven by commercial interests rather than science.

SSRIs for Whatever Ails You

When Pfizer and SmithKline Beecham brought Zoloft and Paxil to market in the early 1990s, Eli Lilly's Prozac was solidly entrenched as the drug of choice for depression. These two manufacturers needed to find new patient groups to target, and both companies identified a group of anxiety disorders, which had been described as discrete conditions for the first time in *DSM-III*, as ripe with possibility.

Psychoanalysts had long understood anxiety neurosis to be a psychological problem that could best be treated through a talking cure. However, for Spitzer and his colleagues, it was not a diagnosis that fit the disease model they were trying to develop for *DSM-III*. In addition, neurosis was a Freudian term, which was the very diagnostic approach that Spitzer was eager to retire. New diagnostic categories needed to be created that had greater specificity than "anxiety neurosis," and the *DSM-III* task force did so by isolating symptoms that were regularly experienced by anxious people, and turning those symptoms into discrete disorders. In *DSM-III*, patients could now be diagnosed with agoraphobia, panic disorder, post-traumatic stress disorder (PTSD), social phobia, obsessive-compulsive disorder, and general anxiety disorder, with each diagnosis said to be distinct from the other.

The surprising element of this categorization effort, given that anxiety was such a common complaint of talk-therapy patients, was that the APA, when it reported the prevalence of these new disorders in *DSM-III*, did not find them—other than panic disorder—to be particularly common. Panic disorder was seen as the principal illness that afflicted anxious people, with *DSM-III* stating that its prevalence was "apparently common." Agoraphobia was said to affect 0.5 percent of the population, and the other anxiety diagnoses were described as "relatively rare" or as having an unknown prevalence, which reflected that they were new constructs.[41]

Following the publication of *DSM-III*, Upjohn moved to get Xanax approved for panic disorder. However, the pharmaceutical industry didn't pay much attention to the other diagnoses, as they were too new or too small to warrant an effort to get FDA approval for that particular disorder. Psychiatrists could always prescribe a benzodiazepine for PTSD or other symptoms of anxiety. But when the NIMH conducted its National Comorbidity Survey in the early 1990s, it produced results that made industry sit up and take notice: 17 percent of the American population were said to suffer from a diagnosable anxiety disorder each year. PTSD, social phobia, and general anxiety disorder were now understood to be quite common. The makers of SSRIs also found reason to be newly interested in panic disorder too, since alprazolam was no longer under patent protection and thus not so heavily marketed by Upjohn.[42]

As Pifzer and SmithKline Beecham sought to sell their SSRIs for these disorders, they recruited academic psychiatrists who had helped formulate the diagnostic criteria for the new disorders to be their "thought leaders."

These psychiatrists were then involved in every step of the process for developing the drugs: they served on advisory boards that designed the clinical trials; they often conducted the trials; they authored reports on the trials; they spoke at industry-funded symposiums at the APA's annual meetings; they served as the experts for media stories about the newly recognized illnesses; and they served on speakers' bureaus. The industry's utilization of academic psychiatrists for "selling the disease" and the drug can be clearly seen in internal SmithKline documents (which were made public during subsequent legal proceedings), and through a review of the published articles by SmithKline's thought leaders.

In November 1993—a year after the FDA approved Paxil for depression—SmithKline convened a meeting of its advisory board in Palm Beach, Florida. The planning document for the meeting identified ten advisory board members (out of 20), three of whom were currently serving on the *DSM-IV* task force: James Ballenger, David Dunner, and Robert Hirschfeld.[43] The board members were flown first class to Palm Beach, where they stayed at the Ritz Carlton Hotel, and each was paid $2,500 to $5,000 for attending the weekend meeting. The advisory board was chaired by Charles Nemeroff, head of the psychiatry department at Emory University.

According to the planning document, the advisory board at the Palm Beach meeting was expected to provide "helpful ideas and advice on issues faced by the PAXIL Marketing Team." Over the course of the two days, the advisory board would discuss the "strengths" and "deficits" of the competitor's SSRIs, and discuss how to educate primary care physicians so they would prescribe Paxil (instead of Prozac or Zoloft). They would also talk about ways to "generate information for use in promotion" of Paxil, and identify "four or five patient subtypes for treatment with Paxil." This last workshop, which was focused on expanding Paxil's market, was led by Ballenger, chairman of the psychiatry department at Medical University of South Carolina. At the end of the meeting, the advisory board would make recommendations for "future studies" that were necessary to support this expanded use of Paxil.

For SmithKline, Ballenger was a good choice to lead a discussion on future studies related to anxiety disorders. He had helped lead Upjohn's trials of alprazolam for panic disorder in the early 1980s, and he had served on the *DSM-IIIR* and *DSM-IV* work panels for anxiety disorders. In the wake of the Palm Beach meeting, Ballenger became the head of the Paroxetine Panic Disorder Study Group for SmithKline Beecham, and in 1996, this group provided the study data that led the FDA to approve Paxil for this disorder. In a ten-week trial involving 278 patients with panic disorder, paroxetine proved "superior to placebo across the majority of outcome measures." Ballenger and his coauthors concluded that "these data support the use of paroxetine for the short-term treatment of panic disorder and its long-term management."[44]

David Sheehan, a professor of psychiatry at the University of South Florida, was another member of the company's advisory board. In 1996, he published a paper on "The role of SSRIs in panic disorder," and at a SmithKline-funded symposium at the APA's annual meeting in 1998, he summed up what was now known about the disease.[45] Panic disorder, Sheehan told his peers, "is a common, chronic, disabling condition with a high-suicide rate." The annual prevalence rate for this disorder was 3 percent (up from 1.3 percent in a 1980s survey). SSRIs, Sheehan concluded, are the "first-line agents for people with panic disorders... patients who have more than two episodes should be maintained on medications indefinitely."[46] The symposium papers were published in a 1999 *Journal of Clinical Psychiatry* supplement.

One anxiety disorder down; three more to go. Next up on SmithKline's agenda was social anxiety disorder (SAD). This was a disorder described as "relatively rare" in *DSM-III*, but that began to change in 1985, when Columbia University psychiatrists Michael Liebowitz and Jack Gorman published a paper on social phobia, calling it a "neglected anxiety disorder."[47] Two years later, Liebowitz served on a *DSM-III-R* work panel that markedly loosened the criteria for diagnosing this disorder. People no longer had to have a "compelling desire to avoid" certain social situations, such as giving a public speech; instead, the diagnosis now only required that the situation caused "marked distress."[48] Partly as a result of this expanded definition, social anxiety disorder soon turned into one of the most "prevalent mental disorders in the United States," behind only major depression and alcohol dependence, with an annual prevalence of 8 percent.[49] SmithKline hired Liebowitz, who had been on the *DSM-IV* task force, as a consultant. Liebowitz then helped lead a trial of paroxetine for SAD, reporting that in an 11-week study, the drug treatment produced "substantial and clinically meaningful reductions in symptoms and disability."[50,51] Then came the selling of the disease. In 1998, SmithKline paid for a symposium on anxiety disorders at the APA's annual meeting, where Liebowitz told the audience that "if left untreated, social anxiety disorder is often a chronic disease and usually does not resolve spontaneously." SSRIs, Liebowitz concluded, should be considered a "first-line therapy" for SAD.[52] The FDA approved Paxil for SAD in 1999.

As SmithKline developed Paxil as a treatment for PTSD, it turned to Jonathan Davidson, a professor of psychiatry at Duke University, to serve as a primary thought leader. Davidson, like Ballenger, had been one of the six members of the *DSM-IV* work panel for anxiety disorders. Like many of his colleagues, Davidson served as an advisor, consultant, and speaker to a number of pharmaceutical companies, including Pfizer and GlaxoSmithKline (SmithKline Beecham merged with Glaxo Wellcome in 2000).[53] In his presentations at industry-funded symposiums and in his articles on PTSD, Davidson explained how PTSD was a "major health concern," with a "lifetime prevalence" of around 9 percent. PTSD, he wrote, was "poorly recognized," even though it should be "viewed as among the most serious of all psychiatric disorders."[54] The efficacy of several SSRIs had been demonstrated in clinical trials, he said, concluding in a 2003 paper that "paroxetine is especially well-studied in this regard, with demonstrated efficacy in men and women, in both short-term and long-term studies, and in combat veterans and civilians."[55] The FDA approved Paxil as a treatment for PTSD in 2001.

GlaxoSmithKline relied on both Ballenger and Davidson as it marketed Paxil as a treatment for general anxiety disorder (GAD). In *DSM-IIIR*, GAD was said to be "not common," but, by the year 2000, Ballenger and Davidson were promoting a new understanding. That year, they led a meeting—which was funded by SmithKline Beecham—of the "International Consensus Group on Depression and Anxiety." The group's "consensus statement" declared that GAD "is a chronic, prevalent, and disabling anxiety disorder." Patients diagnosed with GAD could "be described as having periods of worsening disease alternating with periods of remission." Although benzodiazepines had long been prescribed for anxiety disorders, these drugs, because of their addictive properties, could "be a problem in the long term," and thus SSRIs and other antidepressants, which had proven to be effective in clinical studies, should be the first-line therapy for GAD. The group's consensus statement, with Ballenger listed as first author and Davidson as second author, appeared in the *Journal of Clinical Psychiatry* in 2001, the same year that the FDA approved Paxil as a treatment for GAD.[56]

From the initial advisory board meeting in 1993, it had taken SmithKline eight years to get Paxil approved for the four anxiety disorders (Table 6.1), and thus reposition its drug as the "anxiolytic antidepressant." Three of the four—PTSD, social phobia, and general anxiety disorder—had been seen as uncommon or even rare when they were first identified as discrete illnesses in *DSM-III*, but were now seen as common, disabling illnesses that often required lifelong treatment. The anxious person who had once turned to a psychiatrist or counselor for talk therapy was now someone with a medical problem that could be successfully treated with an SSRI. That was the story newly told by psychiatry's thought leaders who were experts in anxiety disorders, and after SmithKline obtained approval for the four diagnoses, the next phase in its marketing plan kicked into action: the flooding of medical journals with ghostwritten papers, and the hiring of academic and community psychiatrists to serve on its speakers' bureau.

In the United Kingdom, SmithKline hired a firm called The Medicine Group to ghostwrite papers. In a fax to SmithKline dated January 21, 1999, The Medicine Group detailed its plan to write five "pharmacy review" articles. One of the suggested articles was "Pharamacoeconomics of Depression and Anxiety Disorders," which identified Jonathan Davidson as one of the "proposed authors." For an article titled "Paroxetine—Efficacy across the Spectrum," the firm identified Columbia University psychiatrist Jack Gorman as a "proposed author." The Medicine Group was eager to put pen to paper: "Is it possible for us to make a start on any of these articles?" it asked.[57] A year later, the British firm provided SmithKline

Table 6.1   From DSM IV to the market: The selling of Paxil

Investigator	On DSM IV Work Group	At SmithKline Advisory Board Meeting in 1993	Authored Article on Efficacy or Safety of Paxil	Consultant or Advisor to GlaxoSmithKline, or on Speakers' Bureau
James Ballenger	Anxiety	Yes	For panic disorder/GAD	Yes
David Dunner	Mood Disorders	Yes	For depression	Yes
Robert Hirschfeld	Personality Disorders	Yes		Yes
Michael Liebowitz	Anxiety	Yes	For social anxiety disorder	Yes
Jonathan Davidson	Anxiety		PTSD/GAD	Yes

with an update on its "Paxil Publication Plan," the report detailing the progress of the many articles moving through this ghostwriting process. The report noted whether SmithKline had approved the manuscript, and for manuscripts that had passed that hurdle, it named the psychiatrists that had agreed to serve as authors, and how it was now "incorporating author comments to manuscript." The article on "Paroxetine—Efficacy across the Spectrum," had been written, but The Medicine Group was still "awaiting manuscript approval" from SmithKline, and as such, the company had yet to sign up a psychiatrist to serve as the author for the article.[58]

In the United States, SmithKline hired Scientific Therapeutics Information (STI) to ghostwrite articles. One of STI's projects was to write the articles that appeared in a 2003 supplement of the *Psychopharmacology Bulletin*, which was titled "Advancing the Treatment of Mood and Anxiety Disorders: The First 10 Years' Experience with Paroxetine."[59] STI, when it sent Yale Medical School psychiatrist Kimberly Yonkers a first draft of the article she was authoring, helpfully noted that STI's cover page on the manuscript was "to be removed before submission" by Yonkers to the journal.[60] The supplement featured 16 papers on the merits of paroxetine, with Davidson, Sheehan, Nemeroff, and other SmithKline "thought leaders" among the authors. Nemeroff, who had chaired the company's advisory board, "authored" the introduction to the supplement, and in it, he acknowledged that STI had assisted in preparing the manuscripts.[61]

When SmithKline merged with Glaxo Wellcome in 2000, the new company—GlaxoSmithKline (GSK)—created a speaker's bureau called PsychNet. This program, noted a company manual, recruited "key influential psychiatrists and primary care physicians" in each region of the country, with the purpose of developing them "into knowledgeable and engaging speakers on Paxil and its effective treatment on mood and anxiety disorders."[62] At weekend training sessions, they were trained on "a total of five (Powerpoint) presentations," one of which was titled "Contemporary Issues in Anxiety Spectrum Disorders." This presentation, a PsychNet document noted, "reviews the prevalence of anxiety spectrum disorders, current treatment options, and the efficacy of Paxil to treat anxiety." The slides told of how 25 percent of all American adults suffered from an anxiety disorder at some point in their lives.[63]

The PsychNet "faculty" was composed of 65 physicians (mostly psychiatrists), each of whom was expected to give 4 to 15 talks per year. Each speaker was paid $1,000 to come to the training weekend, and $2,500 for every community talk. The faculty included a number of well-known academic psychiatrists, including Sheehan (who helped train the faculty, and served on the PsychNet's Scientific Advisory Board), John Zajecka, Michael Hirsch, and Paul Keck.[64] The speakers had to sign a confidentiality agreement, which precluded them from disclosing "information about the PsychNet program, including compensation and content of materials." The honorarium paid to these "key opinion leaders," noted one GSK document, "is based on the speaker delivering a promotional *Paxil* presentation."

GSK's star speaker was Charles Nemeroff. Even though he was chair of the psychiatry department at Emory University, he still managed to give more than 50 talks in 2000. GSK paid him $2,500 and up per talk, and at least $5,000 for attending Paxil advisory board meetings. From 2000 to 2004, GSK paid him $849,413 for his services.[65]

All of this proved quite profitable for GSK. In 2002, Paxil became the best-selling antidepressant in the world. It was now touted as an "anti-anxiolytic drug," and, as can be seen in this review, it was the APA and academic psychiatry that enabled this commercial success. Prior to *DSM-III*, anxiety was seen as arising from psychological stresses, and those struggling with anxiety often sought out talk therapy. If they took a drug, a benzodiazepine was the "minor tranquilizer" of choice. But then the APA reconceptualized anxiety as a set of discrete illnesses in 1980, and over the next two decades, American psychiatry converted anxious feelings and behaviors into medical diseases, which were understood to be quite prevalent and were best treated with an SSRI.

The question of whether this commercial process provided the public with a medical benefit, with the SSRIs an effective treatment for the various anxiety conditions, is difficult to assess. Many psychiatrists believe that SSRIs are effective as anxiolytics. However, as this market was built, there was an absence of independent, noncommercial research conducted to study the effectiveness of SSRIs for the various anxiety problems. The academic researchers who led the trials and spoke about the validity of the disorders were paid to be the companies' thought leaders, and the medical literature was contaminated by ghostwritten papers, which universally concluded that the SSRIs were safe and effective. Unfortunately, that doesn't provide a scientific record of efficacy.

SSRIs for Children Too

The selling of SSRIs as a treatment for mild forms of depression and anxiety basically turned them into pills for emotional distress of all types. In the 1990s, American psychiatry, in collaboration with industry, identified one other potential market for these drugs: children and adolescents.

As was the case with adult depression, this market—from a diagnostic perspective—was dependent on a break with thinking from the past. Prior to *DSM-III*, moodiness was seen as an ordinary part of growing up. Depression was a disorder that afflicted mostly the middle-aged and elderly. However, once the APA redefined depression in 1980, such that the diagnosis was now divorced from life events (except for bereavement), the thought that moodiness might be "normal" was no longer relevant. The only question was whether youth experienced the same symptoms that led to a diagnosis of a depressive disorder in adults. If so, children and adolescents could be said to suffer from major depression, dysthymic disorder, or depressive disorder not otherwise specified (the latter a diagnosis for "minor depression"). The APA also decided that "irritability" should be seen as a core symptom of depression in children and adolescents, which, as Harvard Medical School psychiatrist Ronald Kessler noted, further expanded the population of youth that could be diagnosed with depression.[66]

Based on the *DSM-III* and *DSM-IV* criteria, epidemiological studies found that depression was quite common in adolescents, and occurred with some frequency in grade-school children too. However, the researchers differed in their reports on precisely how common it was; their studies produced one-year prevalence rates ranging from 2 to 50 percent.[67] Perhaps the most authoritative study was the National Comorbidity Survey Replication funded by the NIMH, which found that based on *DSM-IV* criteria, nearly 12 percent of teenagers experienced a bout of major depression or dysthymia by the time they reached age 18.[68] Science, it seemed, had uncovered a hidden epidemic, one that had been missed by psychiatrists prior to the early 1990s.

"Although epidemiological studies of child and adolescent mood disorders have been carried out for many years, progress long was hampered by two misconceptions: that mood disorders are rare before adulthood and that mood disturbance is a normative and self-limiting aspect of child and adolescent development," Kessler wrote. "Research now makes it clear that neither of these beliefs is true."[69]

With this disease in youth having been newly discovered, academic psychiatrists soon had other news to report: pediatric depression regularly went undiagnosed and untreated. In 1991, Martin Keller, chairman of the psychiatry department at Brown University, reported that in a study of 275 children who came from families with a history of depression, 38 were suffering from this illness and yet none were being treated with an antidepressant.[70] His results revealed a major unmet health need, and on

March 19, 1993, Keller wrote SmithKline Beecham, urging that it fund a study of Paxil for adolescent depression. "My research group discovered underrecognition of adolescents with serious affective illness," he stated, and then he described the discoveries—and the scientific glory—that lay ahead for the company and those who would conduct a pediatric study of Paxil:

> You and your colleagues at SmithKline Beecham will be part of the world's largest and most comprehensive database on adolescent depression...I am confident that the findings generated from this timely research program will result in between 25–40 manuscripts in the leading peer-reviewed journals, such as *New England Journal of Medicine, Journal of the American Medical Association, Journal of Affective Disorders, Archives of General Psychiatry, American Journal of Psychiatry,* and the *Journal of the American Academy of Child and Adolescent Psychiatry.* Moreover, every relevant national and international meeting and workshop held during the next decade will ask to have these data presented. Newly written or edited textbooks will have chapters including findings from this research, and all medical students, interns, residents, fellows and health care professionals will have these data as the latest in the field. In the hands of the Principal Investigators running this program, it is a sure bet that this research will become the world's definitive database on adolescent depression.[71]

Keller was not the only academic psychiatrist eager to test SSRIs in youth. In the early 1990s, Graham Emslie, a professor of psychiatry at the University of Texas Southwestern Medical Center, obtained an NIMH grant to conduct a pediatric trial of fluoxetine, and after that study was underway, Eli Lilly provided him with funds to conduct a larger trial of Prozac. SmithKline agreed to fund Keller's study of Paxil, and soon Pfizer was running pediatric trials of Zoloft. Yet, even before the results from the pediatric trials were reported, many psychiatrists—convinced of the efficacy of SSRIs for adult depression—began prescribing them to children and adolescents. "Until about 15 years ago, no one thought children could suffer depression," the *New York Times* reported in 1997. "Now experts estimate it afflicts about four million American children, or five percent." *The Times* reported that Harold Koplewicz, professor of clinical psychiatry at New York University School of Medicine, "had dozens of children on the new antidepressants" and that these drugs could be "life-saving" for depressed children. "This is no Prozac party," Koplewicz told the *Times.* "These drugs only work if you need them."[72]

In the late 1990s, the results from the pediatric trials of SSRIs began to be announced. Emslie was the first to publish. In 1997, he reported that in his NIMH-funded study, 56 percent of those given Prozac were rated as much or very much improved, compared to 33 percent on placebo.[73] The following year, Karen Wagner, director of child and adolescent psychiatry at the University of Texas, presented a conference paper on the results from the Keller study: "We can say that paroxetine has both efficacy and safety data for treating depression in adolescents," she announced.[74]

The Keller group published their positive findings in 2001. Emslie reported positive results from an Eli Lilly–funded study of fluoxetine in 2002. Pfizer's investigators announced that Zoloft was an effective treatment for major depression in 2003. Soon Forest Laboratories joined this chorus, detailing the merits of Celexa and Lexapro for treating adolescent depression. In 2004, the Center for Science in the Public Interest identified ten reports of industry-funded placebo-controlled trials of SSRIs for adolescent depression, nine of which had found the treatment to be safe and effective.[75]

As this story of a successful medical treatment for pediatric depression unfolded in the medical literature, pharmaceutical money naturally flowed to the academic psychiatrists who led the trials and promoted the merits of SSRIs for this newly discovered pediatric disorder. In 1997 and 1998, Martin Keller earned nearly $1 million in consulting fees from pharmaceutical companies.[76] Karen Wagner, who participated in the testing of several SSRIs for pediatric depression, served as a consultant to 12 drug companies and on the speaker's bureau of seven; from 2000 to 2005, one of the seven companies, GlaxoSmithKline, paid her $160,402 for her services.[77] Forest Laboratories made Jeffrey Bostic, a psychiatrist at Massachusetts General Hospital in Boston, its "star" spokesman, paying him more than $750,000 from 1999 to 2006 to promote the prescribing of Celexa and Lexapro to children and adolescents.[78]

With the scientific literature telling of a helpful medical treatment, and experts in child psychiatry announcing this finding at industry-funded dinners and conferences, the prescribing of SSRIs to youth became ever more commonplace. By 2002, one in every 40 children of school age in the United States was taking an antidepressant.[79]

However, during this period (mid 1990s to 2004), there was a recurring murmur of dissent in the public discourse related to the use of SSRIs in children, which arose from anecdotal accounts of youth killing themselves after being prescribed an SSRI. Did these drugs increase the risk of suicide in adolescents? At last, that murmur rose to such a level that the FDA was forced to convene a public hearing on this question, and in preparation of that 2004 meeting, the FDA dropped a bombshell. Not only did the trial data show that the SSRIs nearly doubled the risk of suicidal ideation in adolescents, 12 of 15 pediatric trials of SSRIs had failed to show efficacy. The FDA had rejected six of the seven applications from SSRI manufacturers seeking pediatric labeling of their drugs, and that was because the studies hadn't found them effective in this patient population. "These are sobering findings," confessed the FDA's Thomas Laughren.[80]

At that moment, the only conclusion to be drawn, which was stunning, was that the many reports published in the medical literature, which had been authored by leading figures in American psychiatry, presented a false story about the safety and efficacy of the SSRIs for depression in adolescents. The data showed harm done, at least for most of the SSRIs, while the medical literature told of a helpful treatment. The Medicines and Healthcare Regulatory Agency in the United Kingdom, now that it had an opportunity to review the trial data, essentially banned the use of SSRIs, except for fluoxetine, in patients under 18 years old. The editors at *Lancet*, a prestigious medical journal published in the United Kingdom, provided a moral context for the entire tawdry affair: "The story of research into selective serotonin reuptake inhibitor use in childhood depression is one of confusion, manipulation, and institutional failure." The fact that psychiatrists at leading medical schools in the United States had participated in this scientific fraud constituted an "abuse of the trust patients place in their physicians," the *Lancet* concluded.[81]

Since that time, a number of scholars have published detailed critiques of the trials, relying both on the FDA's assessments of the pediatric studies and documents that were disclosed as the result of various legal actions against GSK and Forest Laboratories. Their work now provides a clear picture of the methods that were used to turn negative findings—or in the case of Prozac, marginal results—into published articles that told of how the drugs were safe and effective treatments for adolescent depression.

While the FDA did approve fluoxetine for use in children, both of the trials conducted by Emslie relied on a design that washed out initial placebo responders, which could be expected to dampen the placebo response. The study funded by Eli Lilly added a second design element: those who initially responded poorly to Prozac were also washed out, and thus their poor outcomes also didn't count in the final results. As Jonathan Leo, editor of the journal *Ethical Human Psychology and Psychiatry* explained in a 2006 article, "before the study even started, there was a mechanism in place to maximize any difference between the drug and placebo groups— the placebo group was preselected for *nonresponders*, while the drug group was selected for *responders*."[82] Yet, even given that biased design, Prozac still failed to show efficacy on both of its primary outcome measures. There were no significant differences in rates of remission or recovery, the patients' self-ratings of depression, the parents' ratings of symptoms, global psychiatric symptoms, and global functioning. The evidence of a drug benefit came solely from clinician ratings on "selected measures" that were not the primary endpoints designated in the protocol.[83] "One could argue that this post hoc choice of primary outcome is inappropriate," noted FDA reviewer Russell Katz.[84]

SmithKline Beecham conducted three trials of Paxil for pediatric depression, and as a 2011 legal complaint filed by the US government against GlaxoSmithKline noted, all three "failed to demonstrate Paxil's effectiveness while raising concerns regarding an increased risk of suicide among such patients."[85] Keller's study, known as study 329, was one of the three, and in it, Paxil failed to show efficacy on the two primary outcome measures and on the six secondary outcome measures specified in the protocol. Faced with such consistently negative findings, SmithKline privately concluded that it needed to "effectively manage the dissemination of these [trial] data to minimize any potential negative commercial impact...it would be commercially unacceptable to include a statement that efficacy had not been demonstrated, as this would undermine the profile of paroxetine."[86] A data-mining exercise provided the company a way out. SmithKline sorted through the data trying to find an outcome measure that showed Paxil to be significantly better than placebo, and after 19 such possibilities were analyzed, they found four that fit the bill, which provided Keller and his colleagues with a way to report a positive efficacy result.[87]

The safety data also needed a good scrubbing before it could be published. Ten of the 93 teenagers treated with Paxil in study 329 had self-harmed or reported emergent suicidal ideas, compared to only one of 87 in the placebo group.[88] Eleven in the Paxil group had suffered a "serious adverse event" versus two in the placebo group, and as an early analysis noted, the serious adverse events in the Paxil group included "worsening depression, emotional lability, headache, and hostility" that "were considered related or possibly related to treatment."[89] In the published report, these data basically disappeared. The ten instances of self-harm and suicidal ideas in the Paxil group turned into six cases of "emotional lability," and of the 11 serious adverse events in the Paxil group, Keller and colleagues wrote that "only headache (one patient) was considered by the treating investigator to be related to paroxetine treatment."[90] With the safety data recategorized in this way, Keller and his colleagues concluded that "Paroxetine is generally well tolerated and effective for major depression in adolescents." The authors of that article included 16 academic psychiatrists, who were on faculty at ten US medical schools.

As for the pediatric trials of Zoloft, two studies failed to show a statistically significant benefit for the drug. However, when the data from the two failed studies were pooled, it produced a finding that Zoloft provided a statistically significant benefit compared to placebo, as 69 percent of the patients were considered "responders," versus 59 percent of the placebo patients.[91] That was the finding published in 2003 by the Sertraline Pediatric Depression Group, with Wagner as the lead author. However, the paper did not mention that each of the individual studies, when analyzed alone, had not shown the drug to be effective. As for the drug's safety, at least five in the Zoloft group either attempted suicide or experienced suicidal ideas, versus two in the placebo group, but this difference was dismissed as statistically insignificant in the published report.[92] Wagner and colleagues concluded that Zoloft "is an effective and well-tolerated short-term treatment for children and adolescents with major depressive disorders." But regulatory authorities in the United States and the United Kingdom concluded otherwise, with the UK reviewers focusing on the fact that the submitted data "yielded two negative trials."[93]

In the critiques of the pediatric trials for these three SSRIs, it is easy to see that the scientific sins were many. Indeed, many scholars studying financial conflicts of interest in medicine have focused on the pediatric trials of SSRIs as a "worst case" example. In an article titled "One Flew Over the Conflict of Interest Nest," Irish psychiatrist David Healy summed it up this way:

> The difficulties [arising from financial conflicts of interest] are best symbolized by the case of the pediatric trials of selective serotonin reuptake inhibitors, where we have the greatest known divide in medicine between the raw data on the issue on the one side and the published accounts purporting to represent the data on the other. The data can now be seen to indicate that the drugs do not convincingly work and are hazardous, but prior to the release of the data the scientific literature universally portrayed these agents as safe and effective.[94]

The New Bipolar

Although it is often said that bipolar disorder is a modern name for what Emil Kraepelin called manic-depressive illness, that isn't quite accurate, for the bipolar diagnosis represents a subset of patients from Kraepelin's manic-depressive group. Kraepelin, in his study of psychotic patients, found that those who presented with mania, depression, or episodes of both appeared to share a similar outcome: they could expect to recover from their initial bout of mania or depression, and their illness usually ran an episodic course. Kraepelin concluded that mania and depression were different symptoms of a single underlying disease, and thus he grouped these patients into one diagnostic category.

The cleaving of manic-depressive illness into separate unipolar and bipolar diagnoses arose from the studies of a German psychiatrist, Karl Leonhard, who reported in 1957 that there was a stronger familial link in mania than in depression, and that onset was earlier in the manic patients too. His findings suggested that mania was different in kind from depression. He called the manic group "bipolar," a term that reflected the fact that many of the manic patients subsequently suffered from depressive episodes. Once this distinction was proposed, other researchers reviewed the patient records of manic depressive patients from the first half of the twentieth century to assess the prevalence of "bipolar" disorder, and they concluded that such patients comprised 10 to 50 percent of the larger manic-depressive group. At least in the United States at that time, manic depressive illness was infrequently diagnosed, and thus the prevalence of the bipolar subtype was quite small—there were perhaps 12,750 people hospitalized with bipolar illness in 1955, or about one in every 13,000 people.[95]

In 1969, George Winokur at Washington University in St. Louis treated unipolar depression and bipolar illness as distinct entities in his textbook *Manic Depressive Illness.* A year later, the FDA approved lithium as a treatment for mania. At that point, the diagnosis of manic depressive illness began to be made more frequently, with lithium seen as the treatment for the bipolar form. In *DSM-III,* bipolar illness was identified as a discrete disorder for the first time, with a lifetime prevalence said to range from 0.4 percent to 1.2 percent of all adults.[96]

This was a marked increase in prevalence from the pre–lithium days. At the same time, *DSM-III* opened the door to expanding the bipolar diagnosis to a much broader group of patients. In 1977, Hagop Akiskal, a professor of psychiatry at the University of California San Diego, identified a group of patients who suffered from periodic mood disturbances, characterized by bouts of depression and hypomania, only these disturbances were below the threshold for making a diagnosis of either major depression or mania.[97] *DSM-III* incorporated his findings, stating that such patients had cyclothymic disorder, which could be seen as a mild form of bipolar illness.

In the following years, Akiskal promoted the idea of a "bipolar spectrum," with full-blown affective psychosis at one end and milder mood fluctuations at the other. He and a handful of others published regularly on this topic, and *DSM-IV* reflected this research, listing four types of bipolar disorders. There was bipolar 1, characterized by episodes of major depression and mania; bipolar II, characterized by major depression but with the mania component downgraded to an episode of hypomania consisting of at least four days of elevated, expansive or irritable mood; cyclothymic disorder, for patients whose recurring mood symptoms weren't severe enough to meet criteria for either major depression or a manic episode; and "bipolar disorder not otherwise specified," for those who showed "bipolar features that do not meet criteria" for any of the other three bipolar diagnoses. The authors of *DSM-IV* stated that as many as 3 percent of adults suffered at some point in their lives from one of the bipolar disorders (a prevalence rate that excluded the bipolar not-otherwise-specified category.)[98]

In 1996, Akiskal became editor-in-chief of the *Journal of Affective Disorders,* and from that pulpit, he continued to publish his own research and studies by others that further expanded the concept of bipolar disorder. In a 1996 paper titled "The Prevalent Clinical Spectrum of Bipolar Disorders: Beyond DSM IV," he argued that the diagnostic criteria for bipolar disorders in *DSM-IV* were too narrow, and thus "would deprive many patients with lifelong temperamental dysregulation and depressive episodes of the benefits of mood-regulating agents."[99] The "soft bipolar spectrum" needed to be further elucidated, he argued, and he subsequently set forth criteria for diagnosing bipolar III and bipolar IV, the last category for people who experienced "subthreshold hypomanic traits rather than episodes" of elevated mood.[100] In 2000, he reported that when the milder forms of bipolar were included, the true prevalence of the illness was "at least 5 percent."[101] Three years later, he raised this prevalence rate to 6.4 percent.[102] Finally, in 2007—and by this time he had authored or coauthored nearly 300 papers on mood disorders—he concluded that "one of ten individuals in the community either has bipolar disorder, or is at risk for it."[103]

As this new understanding of bipolar was constructed, drug companies, starting in the 1990s, moved to capitalize on this new opportunity. Lithium had been known as an anti-manic drug, but now the drug companies and researchers began talking about "mood stabilizers" as the preferred treatment for bipolar disorder. In 1995, Abbott got an older antiseizure drug, Depakote, approved for use in bipolar patients, and then the makers of atypical antipsychotics rushed to get their drugs approved for bipolar disorder. As they did so, they hired Akiskal to serve as one of their principal thought leaders. In 2010, he disclosed that he had served on speakers or advisory boards for GlaxoSmithKline (Lamictal); Abbot (Depakote); Eli Lilly (Zyprexa); Janssen (Risperidone); AstraZeneca (Seroquel); and Bristol Myers Squibb (Abilify).[104]

Internal Eli Lilly documents, which were made public during legal proceedings, illuminate how that company sought to rebrand its drug as a mood stabilizer, and the importance of "thought leaders" to this effort. In 1997, a year after the FDA had approved Zyprexa as an antipsychotic, the "Zyprexa Product Team" forecast that sales could increase more than fourfold if the drug could be repositioned as a "Depakote-like . . . MOOD-STABILIZER" instead of a "Risperdal-like Antipsychotic."[105] In order for this to happen, the company needed to develop a "scientific research and publication plan that enhances credibility of the new brand positioning and enables the achievement of the ideal positioning." The company, in a document on "influencing key players," noted that the company's top thought leaders "are well respected and acknowledged by their peers . . . [they] influence the thinking and treatment practices of their peers . . . and are typically in the academic setting and treat a minimal number of patients, if any . . . and serve on academic advisory boards, providing feedback to the Zyprexa Product and Brand Team."[106]

Eli Lilly obtained FDA approval to market Zyprexa for bipolar disorder in 2000, and soon it became one of the top-selling drugs in the world, regularly generating more than $4 billion in annual revenues. At that time, AstraZeneca was feverishly trying to get its drug, Seroquel, approved as a bipolar medication, and as part of that effort, it conducted a series of interviews of clinical researchers in Europe who had worked as "KOLs" for Eli Lilly, Janssen, and other manufacturers of atypical antipsychotics. In an e-mail dated August 8, 2003 to the Seroquel Global Brand Team, AstraZeneca's "Global Brand Manager" for Seroquel summed up what had been learned about studies that had been initiated by the investigators (as opposed to the company). Eli Lilly, the e-mail stated, offered "significant financial support" to investigators who wanted to conduct trials of

Zyprexa, but the company sought "control of the data in return. They [Eli Lilly] are able to spin the same data in many different ways through an effective publications team. Negative data usually remains well hidden." The same was true of Janssen, the e-mail stated. The company didn't allow investigators, even when they had initiated the studies, to publish findings "without going through Janssen for approval... communication is controlled by Janssen." The KOLs who published favorable results "are well rewarded for their involvement" by Janssen. AstraZeneca's global brand manager then issued his recommendation: their company "should be more creative spinning data" in its published reports, much in the same manner that Eli Lilly had done.[107]

Today, there are a number of examples of how this process led to publications in the medical literature that produced a misleading picture of the efficacy of several of the drugs approved as mood stabilizers. Here are two examples:

- GlaxoSmithKline conducted nine studies of lamotrigine (Lamictal) for treating acute bipolar episodes and for maintenance treatment. The two maintenance studies that were positive were published. A negative study in rapid cycling patients and a negative study in acute bipolar depression were published, but the published articles emphasized positive results on two secondary outcomes as opposed to the negative results on the primary outcome measures (e.g., negative studies were cast as positive ones). There were five more negative studies involving rapid cycling bipolar disorder, acute bipolar depression, and acute mania that went unpublished.[108]

  The disclosure of the negative studies occurred as the result of a lawsuit by New York State against GlaxoSmithKline for its fraudulent marketing of paroxetine in children. Nassir Ghaemi, a professor of psychiatry at Tufts University, published an article on this larger data pool once it was made public, and he concluded that while lamotrigine was "reasonably effective" as a maintenance therapy, it "is proven ineffective in acute mania, rapid cycling disorder, and acute bipolar depression."[109]

- In May 1996, Parke-Davis, the maker of gabapentin (Neurontin), a drug approved for treating epilepsy, noted that the company had launched an "exploratory study in bipolar disorder... the results, if positive will be publicized in medical congresses and published in peer-reviewed journals."[110] Over the next seven years, there were forty *open-label* studies published in medical journals stating that it was effective in treating bipolar disorder. However, the company's first randomized, double-blind study of Neurontin for this condition, which was completed by July 18, 1997, showed that the Neurontin patients did "significantly worse" than those taking placebo.[111] Three other randomized trials also failed to show that gabapentin provided a benefit over placebo. And while the results from the first randomized study were eventually published in 2000, from 1997 to that publication date, Parke-Davis relied on its thought leaders and speakers' bureau to market the drug as an off-label treatment for bipolar disorder, with these thought leaders espousing the drug's efficacy. This effort was so successful that Neurontin became one of the leading drug treatments for this condition.

For instance, at the 10th annual US Psychiatric and Mental Health Congress in Orlando, Florida, on November 15, 1997, John Zajecka, an associate professor of psychiatry at Rush University Medical Center in Chicago, presented a seminar titled "New Options for Bipolar Disorders." He told the audience that Neurontin was "generally well tolerated" and that its efficacy was "under investigation," even though by that time Parke-Davis's randomized study had been completed, with the Neurontin patients faring worse than placebo.[112] The company then had its thought leaders, all of whom "held prestigious academic positions," speak at "CME Psychiatry Dinners" in expensive restaurants around the country, with the slide presentation they used titled: "Closing the Psychiatry-Neurology Divide: Emerging Uses of Anticonvulsants." Bipolar disorder was said to be responsive to Neurontin, with one slide stating that it was "indicated for bipolar disorder" (which falsely suggested that it had been approved by the FDA for this purpose).[113] Pfizer, which had purchased Parke-Davis, was subsequently sued for its improper marketing of Neurontin for off-label use, and after reviewing company documents, John Abramson, a physician affiliated with Harvard Medical School who served as an expert witness in one of the cases, concluded that Parke-Davis had made "use of Advisory Boards and thought leaders" to "exploit physicians' trust in the sources from which they typically assimilate new knowledge." In this way, the company got physicians to prescribe Neurontin for bipolar disorder, even though randomized studies had failed to show that the drug was effective for this purpose.

Such is the story of the bipolar boom in adults. The APA and other bipolar experts created an expansive definition of the disorder, such that it could be seen as an ailment that afflicted one in every 16 adults, and pharmaceutical companies then rushed to market their drugs as "mood stabilizers." They recruited academic psychiatrists to serve as their advisors, consultants, and speakers during this process, and, at least from a business perspective, in the end it produced a financial success. The mood stabilizers became best-selling drugs, so much so that the number of prescriptions of atypical antipsychotics for mood disorders soon outstripped the number written for psychotic disorders, helping antipsychotics become the top revenue-generating class of drugs in the United States in 2009, with $14.6 billion in sales.[114]

DSM-5: The Expansion Continues

From an industry perspective, the APA's steady expansion of diagnostic boundaries, starting with *DSM-III* and then continuing with subsequent editions of the DSM, proved to be a remarkable success. Meanwhile, the APA saw this expansion as consonant with its public health mission, with the field now able to diagnose and effectively treat previously unrecognized illnesses. The number of Americans being diagnosed with a psychiatric disorder steadily climbed, which in turn produced a great increase in the sales of psychiatric drugs and psychiatry's influence over American society. Indeed, psychiatry could easily lay claim to being the most influential medical discipline in America today, as it is reshaping our society in profound and far-reaching ways. The fifth edition of the APA's manual, published in 2013, continues this pattern of diagnostic inflation.

As was the case with *DSM-IV*, the majority of those helping to create *DSM-5* had financial ties to industries (Table 6.2). In 2007, the APA required that the *DSM-5* contributors file disclosure statements, which revealed that 69 percent of the task force members, who were responsible for guiding the overall process, had a financial tie to industry (compared to 57 percent of the *DSM-IV* task force). Seventy-five percent of the work panels had a majority of members with ties to industry, and as was true with *DSM-IV*, the most conflicted panels were for disorders typically treated with drugs. Two-thirds of the panel members for mood disorders, 83 percent of the panel for psychotic disorders, and 100 percent of the panel for sleep/wake disorders disclosed they had ties to pharmaceutical companies.[115]

DSM-5 expanded diagnostic boundaries in a number of ways.[116] Here are a few of the most notable changes:

Table 6.2   Financial conflicts of interest among *DSM-IV* and *DSM-5* members

	On *DSM-IV* (%)	On *DSM-5* (%)
Task Force	57	69
Work Panels		
Anxiety	81	57
Eating disorders	83	50
Mood disorders	100	67
Sleep disorders	50	100
Schizophrenia/psychotic disorders	100	83

Source: L. Cosgrove. "A comparison of DSM-IV and DSM-5 panel members' financial associations with industry." *PLoS Med* 9 (2012): e1001190.

- Premenstrual dysphoric disorder (PMDD) and binge eating disorder were elevated from the appendix in *DSM-IV* to full-fledged disorders. The latter diagnosis can be given to someone who overeats—in a binge manner—at least once a week for three months, in the absence of bulimia or anorexia.
- The criteria for diagnosing ADHD were revised such that the characteristic symptoms no longer need to be present before age seven. If some of the symptoms used to diagnose ADHD were present before age 12, the diagnosis can now be made in older individuals, which could result in more teenagers and adults being diagnosed with ADHD.
- The bereavement exclusion for diagnosing major depression was eliminated. A person who has suffered the loss of a loved one can now be diagnosed with major depression after two weeks if they meet the usual criteria. *DSM-IV* stated that the diagnosis should not be made unless the depression extended beyond two months.
- Disruptive mood dysregulation disorder is a new diagnosis, intended for children six years and older, who exhibit persistent irritability and have three or more temper tantrums per week. This diagnosis is supposed to be distinct from oppositional defiant disorder, conduct disorder, and bipolar disorder.
- Other new diagnoses include hoarding disorder (a diagnosis for those who have a "persistent difficulty discarding or parting with possessions") and skin-picking disorder.
- For aging baby boomers, *DSM-5* now provides a diagnosis of "mild neurocognitive disorder," characterized by "less disabling syndromes that may nonetheless be the focus of concern and treatment." The 60-year-old who shows signs of forgetfulness will be a candidate for this diagnosis.
- Attenuated psychosis syndrome was put in the appendix as a condition requiring further study; this is a diagnosis given to someone exhibiting "sub-threshold psychotic symptoms."

In sum, *DSM-5*—from a commercial perspective—will provide a diagnostic rationale for treating more children with an antipsychotic; more teenagers and adults with a stimulant; more menstruating women with an SSRI; more people with behavioral quirks with an SSRI (since hoarding and skin picking are seen as obsessive conditions); more grieving individuals with an SSRI; more teenagers and other young adults seen as "prepsychotic" with an antipsychotic; and more forgetful older folks with a drug of a type yet to be determined. This is not to say that this is the APA's intent. The point, as made earlier in this chapter, is that there were economies of influence present during the development of *DSM-5*, and this manual will further enable the APA to play handmaiden to industry.

Indeed, even before *DSM-5* was published, pharmaceutical companies were running trials of their drugs for the new diagnoses. Eli Lilly was testing Cymbalta for bereavement-related depression and binge eating disorder; GlaxoSmithKline was studying Lamictal for binge eating disorder; Shire had trials of Vyvanse underway for binge eating disorder and severe mood dysregulation; Johnson and Johnson was testing Razadyne for mild cognitive impairment; and a handful of other sponsors had drugs in trials for these various diagnoses.[117]

In a study of 13 such trials, researchers identified a web of financial conflicts of interest.[118] In three of the 13, the company testing its drug for the new disorder had paid at least one member of the *DSM-5* work panel that had "decision-making authority" for developing the new diagnosis to serve on its speaker's bureau. Similarly, in three instances, a member of a *DSM-5* work panel that had decision-making authority for creating a new diagnosis was serving as a principal investigator in the trial of a drug for that indication. Finally, of the 41 principal investigators in the 13 trials, at least eight had financial ties other than research funding to the drug companies that were testing their drugs for the new conditions, and four of the eight served on speakers' bureaus for the companies. In short, a number of the members of the *DSM-5* work panels that established the new diagnoses had ties to the drug companies that were now testing their new drugs for those conditions.

Moreover, these trials often promised to effectively extend the patent life of the drug being tested for the new condition, which is of great financial importance to their manufacturers. The patent protection for Eli Lilly's Cymbalta was scheduled to expire in December 2013, and if Eli Lilly could get it newly approved for binge eating disorder and bereavement-related depression, no generic version of the drug could be marketed for this condition until 2016. Five of the 12 members of the *DSM-5* work group responsible for the bereavement-related change had ties to Eli Lilly. Three of the 12 members of the *DSM-5* work group for binge eating had such financial connections to the company.

This expansion of diagnostic categories—and thus commercial opportunities for industry—is happening even as Thomas Insel, director of the NIMH, and many other leading figures in America have publicly stated that the DSM diagnoses "lack validity."[119] Moreover, as was noted in chapter 4, for a diagnostic manual to be useful, it should be reliable too, and the field trials of *DSM-5* revealed that this new manual also failed that test.* Of the 23 diagnoses that were field tested, only five produced a kappa score of .60 or above, which was the level that Spitzer had

---

* The APA declared the field trial results to be fairly "good," but that conclusion was belied by the kappa scores for the 23 diagnoses and previous definitions of what could be considered good, satisfactory, unsatisfactory, etc. The dramatic decline in the kappa scores for schizophrenia, major depressive disorder and major depressive disorder in *DSM 5* compared to *DSM III* is particularly revealing about how poor the scores were for the new manual.

deemed "good" when testing *DSM-III*. Nine of the 23 kappa scores in the *DSM-5* field trials fell between .4 and .59, which by *DSM-III* standards would have been seen as rather poor, and the remaining nine failed to reach the .4 level, which in the past would have been seen as unacceptable. This reliability failure can be seen most notably in the kappa scores for the major disorders: .46 for schizophrenia, .25 for major depressive disorder, and .2 for major anxiety disorders. The last two kappa scores are so low that they tell of diagnostic agreement that is little better than chance.[120]

In addition, the field trials only tested a small number of disorders in *DSM-5*. Since the "reliability of the majority of disorders remains untested," wrote Stijn Vanheule, a professor at Ghent University in Belgium, "the idea that the DSM is a reliable instrument is simply wrong."[121]

Given such results, our society can only conclude this: The APA has provided us with a diagnostic manual that, from a scientific perspective, cannot lay claim to being reliable or valid, which are the twin requirements for a medical manual to be clinically useful. However, as this chapter documents, the manual has proven to benefit the interests of the pharmaceutical industry and the guild interests of the psychiatric profession, and *DSM-5* will continue to nurture those ends.

The Long View

Through the rear-view mirror of history, it is easy to see that the expansion of diagnostic boundaries over the past 35 years, and the corresponding dramatic increase in the prescribing of drugs, could have been predicted in the early 1980s, when psychiatry and industry came to see their interests as closely aligned. This was at a time when academic and industry "partnerships" were being promoted as desirable in all areas of medicine, providing psychiatry with reason to think, even as its relationships with industry deepened, that it was proceeding in a manner that should benefit public health. However, what distinguishes psychiatry from other medical disciplines is that it does not have biological markers for its conditions, which means that it is more vulnerable to commercial influences when making decisions about diagnostic criteria. Given the APA's own guild interests, it is almost inevitable that the APA would steadily expand diagnostic boundaries in a way that served commercial interests.

In a 2011 interview, Robert Spitzer spoke of how the drug companies understood that a new commercial world would unfold with the publication of *DSM-III*, as its new disease model was certain to trigger a "gold rush." The pharmaceutical companies, Spitzer said, "were delighted."[122]

CHAPTER SEVEN

Protecting the Market

> People usually remain unaware that they are acting immorally as the result of a conflict of interest.
>
> —Paul Thagard, 2007[1]

The APA, through its expanded criteria for diagnosing mental disorders, and academic psychiatry, through its legitimization of the expanded boundaries and its reports of the efficacy of drugs for these patient groups, helped create a much larger market for psychiatric medications. Societal spending on psychiatric drugs increased from approximately $800 million in 1987 to more than $35 billion in 2010. But this was a market built on a compromised scientific foundation, and time and again, the field struggled to maintain societal belief in these medications, particularly in the face of study results that raised questions about their merits over longer periods of time—one year and beyond.

The National Institute of Mental Health was typically the funder of this research. As such, the pharmaceutical economy of influence was not directly present in these studies (although the investigators may have had ties to industry). But the guild influence remained, and thus in these trials we can assess most clearly its possible corrupting effect.

Protecting the ADHD Market

As the ADHD market significantly grew during the 1990s, the NIMH conducted a study known as the Multisite Multimodal Treatment Study of Children with ADHD (MTA study). At its outset, the NIMH investigators, led by Peter Jensen, associate director of child and adolescent research at the NIMH, noted that "the long-term efficacy of stimulant medication has not been demonstrated for *any* domain of child functioning."[2] This study, which the NIMH called the "first major clinical trial" it had ever conducted of a "childhood mental disorder," was expected to fill in that evidence gap.

The study randomized 579 children, ages seven to ten years, to four groups: medication management, behavioral therapy, those two treatments combined, and standard community care. In 1999, the MTA investigators announced their results. While all four groups had improved by the end of 14 months, "carefully crafted medication management" had proven to be superior to behavioral treatment in terms of reducing core ADHD symptoms.[3] Psychiatry now had a result that supported longer-term use of the drugs. "Since ADHD is now regarded by most experts as a chronic disorder, ongoing treatment often seems necessary," the investigators wrote.

However, the MTA study was not over. After 14 months, it turned into a naturalistic study. The investigators continued to follow the children, and during this phase, the children were free to seek any treatment they wanted. This enabled the MTA researchers to compare long-term results for children who took stimulants and those who did not, regardless of their initial treatment, and at the end of three years, the data told a different story. At that point, "medication use was a significant marker not of beneficial outcome, but of deterioration," the researchers reported. "That is, participants using medication in the 24-to-36 month period actually showed increased symptomatology during that interval relative to those not taking medication."[4] In addition, the medicated youth had higher delinquency scores at three years, and they were also an inch shorter and weighed six pounds less than the unmedicated group.[5,6]

The six-year results were more of the same. Ongoing medication use was "associated with worse hyperactivity-impulsivity and oppositional defiant disorder symptoms," and with greater "overall functional impairment." Children who had initially received behavioral therapy were also much less likely to be depressed or anxious at the six-year mark. The one exception to these negative outcomes was that medicated youth performed better on a math achievement test.[7]

These results stunned the NIMH investigators. After analyzing the three-year results, they reasoned that "selection bias" might explain this confounding outcome. Those who stayed on the drugs through year three may have had worse ADHD symptoms early in the trial, which would explain why they had fared worse over this longer period. However, when the researchers looked at this possibility they found the opposite to be true: if anything, the off-med group at the end of year three had more severe symptoms early in the trial. The researchers also divided the patients into five subsets, based on different criteria, to see whether there were groups of youth who were doing better on the medication. Once again, "there was no evidence of the superiority of actual use of medication in any quintile."[8]

At that point, the investigators concluded that the worse outcomes for the medicated patients at the end of three years could not be easily explained.

The findings...were not consistent with views and expectations about medication effects held by many investigators and clinicians in the field. That is, long-term benefits from consistent treatment were not documented; selection bias did not account for the loss of relative superiority of medication over time; there was no evidence for "catch up" growth; and early treatment with medication did not protect against later adverse outcomes.

This could be said to be a record of good science. The three-year results confounded expectations, and the authors had published that finding. They had written that medicated children had *deteriorated* in year three compared to the unmedicated children. They reported that outcomes for the medicated children remained worse at year six. The investigators subsequently published an article telling of how the poor outcomes for the medicated youth could not be attributed to "selection bias." The details were all there in the research literature, including the fact that the worse outcomes for the medicated children could not be easily explained away. To complete this record of good science, all the investigators needed to do was clearly communicate that finding to the medical community and to the public. Then the MTA trial would serve as an example of psychiatry putting duty to its patients ahead of its guild interests, and doing so even though Jensen and nine other MTA investigators had extensive financial ties to pharmaceutical companies, including those that manufacture ADHD medications (Table 7.1).

However, while the MTA investigators published the results, they did not *publicize* them, and that critical difference can be easily traced. The story told above had to be dug out from the research literature through a careful reading of the texts. There was no mention of the worse outcomes for the medicated children in any of the paper titles, or in the abstracts, and this finding was not communicated to the public either.

In the abstract of the article reporting the three-year results, the MTA authors wrote that "by thirty-six months, the earlier advantage of having had fourteen months of the medication algorithm was no longer apparent, possibly due to age-related decline in ADHD symptoms, changes in medication management intensity, starting or stopping medications altogether, or other factors not yet evaluated."[9] That conclusion tells of an initial benefit of medication appearing to *wane,* as opposed to a finding, which could be read in the text, of medication being a marker of "deterioration" by the end of that period.

Meanwhile, the NIMH's press release announcing the three-year results reported good news: "Improvement Following ADHD Treatment Sustained in Most Children." The MTA researchers, the release said, were "struck by the remarkable improvement in symptoms and functioning across all treatment groups." Even though continuing medication use "was no longer associated with better outcomes by the third year," the release assured parents that they were not to worry. "Our results suggest

Table 7.1 Financial disclosures of MTA investigators (Number of ties to pharmaceutical firms)

Investigator	Academic Affiliation	Research Funding	Advisory Board	Consultant	Speaker's Bureau
Peter Jensen, M.D.	Columbia Univ.	1		6	6
L. Eugene Arnold, M.D.	Ohio State Univ.	7		5	4
James Swanson, Ph.D.	U California, Irvine	12	11	14	9
Howard Abikoff, Ph.D.	New York Univ.	4		7	3
Laurence Greenhill, M.D.	Columbia Univ.			14[a]	
Lily Hechtman, M.D.	McGill Univ.	5	4		4
Glen Elliott, M.D.	Duke Univ.	4		2	3
Jeffrey Epstein, Ph.D.	U California, Irvine	4	1		2
Jeffrey Newcorn, M.D.	Mt. Sinai Medical School	8	16[b]		7
Timothy Wigal	U California, Irvine	4			2

[a] Research funding and consulting ties disclosed together.
[b] Advisory board and consulting ties disclosed together.
Source: Disclosure statement in B. Molina, "MTA at 8 years." *J Am Acad Child Adolesc Psychiatry* 48 (2009): 484–500.

that medication can make a long-term difference for some children if it's continued with optimal intensity, and not started or added too late in a child's clinical course," Jensen said.[10]

Finally, when the MTA researchers published their six-year and eight-year results (together in one paper), they reported in the abstract that there were no significant differences in outcomes at six years relating to the treatments the children had initially received.[11] The fact that the youth still on medication had worse outcomes on most measures at the end of six years could, once again, be found only through a close reading of the text.★

Although it is now well known within the ADHD research community that the MTA study failed to find a long-term benefit with stimulants, there is almost no commentary in the literature about the finding that medication use was, in fact, associated with *worse* outcomes at three and six years. This might not be so alarming if the MTA long-term results

★ The eight-year outcomes for the medicated and unmedicated children are not clear. A draft of this paper, titled "Author Manuscript," stated that medication use at "later assessments," which would have included the eight-year results, "was associated with worse functioning and more school services." That indicates that the outcomes for the medicated children remained worse at eight years. However, that finding was changed in the published manuscript to this line: "Most associations [between medication use and worse outcomes] were not significant at 8 years." The "Author Manuscript" was accessed through NIH Public Access, on February 15, 2014.

were an anomaly, but they are not. The Western Australian Department of Health, in a study of the ten-year outcomes of children with ADHD, found that the medicated youth were ten times more likely to "perform below their age level" in their school work than the unmedicated ones.[12] In 2013, researchers conducting a long-term study of youth in Quebec found that ADHD medications led to a worsening of educational outcomes, a deterioration in their relationship with parents, and, among the girls, an increased likelihood that they would be diagnosed with other mental and emotional disorders.[13]

Such findings, if properly publicized, would threaten current prescribing practices. But these outcomes remain little known, particularly in the public arena, and in this way, the market for ADHD medications has been protected. Indeed, in 2013, the APA and the American Academy of Child and Adolescent Psychiatry published a brochure titled *ADHD Parents Medication Guide*, which informs parents that ADHD medications work well, and the brochure even identifies the specific study that proves that to be so.

> To help families make important decisions about treatment, the National Institute of Mental Health began a large treatment study in 1992 called the Multimodal Treatment Study of Children with ADHD (Or the MTA study). Data from this 14-month study showed that stimulant medication is most effective in treating the symptoms of ADHD, as long as it is administered in doses adjusted for each child to give the best response—either alone or in combination with behavioral therapy. This is especially true when the medication dosage is regularly monitored and adjusted for each child.[14]

In a sense, that is an accurate statement. But the study didn't end at 14 months, and there is nothing in the guide that tells of the three- and six-year results, which clearly would be information—given the poor outcomes for the medicated youth—that parents would like to know.

Saving the SSRIs

Although the medical journals, during the 1990s, were filled with reports of the efficacy of SSRIs and the other new SSRIs, there was recognition within the research community that such studies didn't capture the efficacy of the drugs in real-world patients, since industry-funded trials excluded those with long-term depression, suicidal thoughts, or with comorbid problems (anxiety, substance abuse, etc.) Various studies found that 60 to 90 percent of "real-world patients" would be ineligible to participate in the industry trials because of the exclusion criteria.[15]

As a result, noted John Rush, a psychiatrist at Texas Southwestern Medical Center in Dallas, "both shorter- and longer-term clinical outcomes of representative outpatients with nonpsychotic major depressive disorder treated in daily practice in either the private or public sectors are yet to be well-defined."[16] With funding from the NIMH, Rush assessed such outcomes. He and his colleagues treated 118 "real world" patients with antidepressants and also provided them with emotional and clinical support "specifically designed to maximize clinical outcomes." This was a well-designed study, and in 2004, Rush reported his results: Only 26 percent of the real-world patients responded to the antidepressant during the first year of treatment (meaning that their symptoms decreased by at least 50 percent on a rating scale), and only about half of that group had a "sustained response." Even more dispiriting, only 6 percent of the patients saw their depression fully remit and stay away during the year-long study. These "findings reveal remarkably low response and remission rates," Rush concluded.[17]

There were other discouraging findings that kept popping up. In a NIH-funded study that compared Zoloft to St. John's wort to placebo, a slightly higher percentage of placebo patients had a "full response" at the end of eight weeks than patients in either of the other two groups.[18] The NIMH also funded a trial that compared Zoloft to Zoloft plus exercise and to exercise alone, and at the end of ten months, 70 percent in the exercise-alone group were well, compared to fewer than 50 percent in either of the groups treated with Zoloft.[19] At least in this study, Zoloft appeared to detract from the benefits of exercise.

All of these studies raised questions about the effectiveness of antidepressants, particularly over longer periods of time. But the number of patients in these studies was fairly small. A more robust study was needed, and around 2000, the NIMH launched what it touted as the "largest and longest study ever done to evaluate depression treatment."[20] Rush was named the lead investigator of the $35 million trial, which was dubbed the "Sequenced treatment alternatives to relieve depression (STAR*D) study."

This study in real-world patients, the NIMH investigators emphasized, was expected to produce results that would guide clinical care in the United States. "Given the dearth of controlled data [in real-world patient groups], results should have substantial public health and scientific significance, since they are obtained in representative participant groups/settings, using clinical management tools that can easily be applied in daily practice," the STAR*D investigators wrote.[21] The results, the NIMH promised, would be "rapidly disseminated."

The Study Protocol

According to the STAR*D protocol, the patients enrolled into the study would need to be at least "moderately depressed," with a score of 14 or greater on the Hamilton Rating Scale of Depression (HAM-D). They would be treated with Celexa at their baseline visit, and then, during the next 12 weeks, they would have five clinical visits. At each one, a coordinator would assess their symptoms using a measurement tool known as the Quick Inventory of Depressive Symptomatology (QIDS-C). The patients would also self-report their symptoms during each visit using this same measuring stick (QIDS-SR). As this was a study meant to mimic real-world care, the physicians would use the QIDS data to help determine whether the Celexa dosage should be altered, and whether to prescribe other "non-study" medications, such as drugs for sleep, anxiety, or for the side effects caused by Celexa. The protocol specifically stated that "research outcomes assessments are not collected at the clinic visits," indicating that the QIDS data would not be used to report study results.[22]

Anybody who dropped out before the end of 12 weeks would be counted as a "non-remitter." At the end of that treatment period, independent "interviewers" would assess the patient's symptoms over the phone using the HAM-D. Remission was defined as a HAM-D score of less than eight. Those who had remitted would be enrolled in a year-long maintenance study, designed to assess whether antidepressants were an effective treatment over this longer period. Patients who hadn't remitted during the initial 12 weeks could select another form of acute care, such as switching to another antidepressant, or adding another antidepressant to Celexa. Patients who failed to remit during this second stage of treatment could then move on to a third phase (where they would be offered a new treatment mix), and those who failed to remit in the third stage would then get one final chance to remit. In each instance, the HAM-D would be used to determine whether a patient was now in remission, and anyone who was in remission would then be entered into the maintenance phase of the study.

The Published Results

In January 2006, the STAR*D investigators reported results from the first stage of treatment. Although 4,041 patients had been enrolled in the study, there were only 2,876 "evaluable" patients. The nonevaluable group was comprised of 607 patients who had a baseline HRSD score of less than 14 and thus weren't eligible for the study; 324 patients who had never been given a baseline HAM-D score; and 234 who failed to return after their initial baseline visit. Slightly less than 28 percent of the evaluable patients remitted during stage one, with their HAM-D scores dropping to seven or less.[23]

While this seemed to be a careful reporting of outcomes, there was one element of the calculation discordant with the protocol. The protocol stated that those who dropped out of the study without taking an exit HAM-D assessment were to be classified "as nonremitters a priori," and thus the 234 patients who had failed to return after their baseline visit, when they were first prescribed Celexa, should have been counted as treatment failures rather than as nonevaluable patients. If the investigators had done this, it would have lowered the reported remission rate to 25 percent.

There was one other oddity in this article. The STAR*D investigators reported a second remission rate using the QIDS-SR scale, without noting that the protocol specifically excluded using it for this purpose. Instead, in this report, the investigators decided that if a dropout, at his or her last clinic visit, had scored as in remission on the QIDS-SR scale, then that person could be added to the pool of QIDS remitters. A treatment failure was transformed into a treatment success in this way. This switch added 152 patients to the QIDS-SR remission list, with the STAR*D investigators reporting a remission rate of 33 percent on this scale (compared to the reported 28 percent using the HAM-D).

This first article had set a precedent for reporting results in ways that didn't follow the protocol and inflated remission rates, and two months later, the STAR*D investigators, while announcing remission rates for the patients who had entered stage two (those who failed to remit on citalopram alone), went a step further in this regard. As had been noted in the January report, there were 931 patients who had baseline HAM-D scores of less than 14 or else didn't have a baseline HAM-D score at all, and thus were not "evaluable" patients. However, the patients in this nonevaluable group continued in the study, and in their March 2006 report, the STAR*D investigators included those ineligible patients in their reported outcomes (tallying up remissions in this second step). But the investigators didn't inform readers of this fact, and thus what this second article did was set the stage for STAR*D investigators to eventually report a cumulative remission rate, at the end of four rounds of treatment, that included 607 patients who hadn't been depressed enough to be eligible for the trial in the first place.[24]

In November 2006, the STAR*D investigators provided a comprehensive report on outcomes, from both the acute and maintenance phases of the study.[25] In this article, there was no reporting of HAM-D scores. Instead, the STAR*D investigators reported remission rates based only on the QIDS-SR scores, with this assessment tool now presented to readers in this way: "The QIDS-SR was not used to make treatment decisions, which minimizes the potential for clinician bias." This was a statement that assured readers they could have confidence in the validity of this tool for assessing outcomes; only readers of the protocol would have known that it was not supposed to have been used for this purpose, and that, in fact, the QIDS data had been used to guide decisions at each treatment visit about whether to prescribe other drugs and alter dosages of the antidepressant.

Having switched to the QIDS-SR scale, the STAR*D investigators reported, in the abstract of this article, that "the overall cumulative remission rate was 67 percent." In the text, they explained how they had arrived at this number. They reported that 50 percent of the patients who had entered the study had remitted by the end of the fourth round of treatment. Then they calculated a "theoretical remission rate." If all of those who had dropped out during the acute phase had stayed in the study through all four levels of treatment, and if the dropouts had remitted at the same rate as those who did stay in all four rounds, then this would have added another 606 remitters to their count, they reasoned.

In sum, over the course of these three papers, the STAR*D investigators had progressively deviated from the protocol in ways that inflated remission rates (Table 7.2). They had excluded 234 patients who had dropped out after their initial baseline visit from their list of evaluable patients, which lowered the number of treatment failures. They had included patients who weren't depressed enough to be eligible for the trial in their reports of remitted patients, which increased the number of

Table 7.2 Deconstructing STAR*D

Deviation from Protocol	Number of Evaluable Patients	Number of Remitted Patients	Remission Rate
No deviation from protocol	3,110	1,192	38%
Excluded 234 patients who had been prescribed Celexa at baseline visit and never returned for a post-baseline visit. These patients should have been counted as treatment failures	2,876	1,192	41%
Added 607 patients who weren't depressed enough to be eligible for the study, and 324 others who lacked a baseline HAM-D score. This added 456 to the list of remitted patients	3,807	1,648	43%
Excluded 136 more patients who were said to have never returned for a post-baseline visit	3,671	1.648	45%
Switched from HAM-D scale to QIDS-SR scale to report outcomes. This added 206 to list of remitted patients	3,671	1,854	51%
Calculated theoretical remission rate based on thought that dropouts would have remitted at same rate as those who stayed in all four stages of acute treatment. This added 606 to list of remitted patients	3,671	2,460	67%

Data for chart is derived from reports by STAR*D investigators. Also see H. Pigott. "STAR*D: A tale and trail of bias." *Ethical Human Psychology and Psychiatry* 13 (2011): 6–28.

treatment successes. They had switched from the HAM-D scale to the QIDS scale to report outcomes, which turned a number of dropouts who should have been counted as treatment failures into remitters. Then they calculated a theoretical remission rate, and when publishing this number in the abstract, they left out the "theoretical" part. Psychologist Ed Pigott, who later published two articles deconstructing the STAR*D study, calculated that if the protocol had been followed, 38 percent (1,192 out of 3,110) would have been reported as remitted during this acute phase of the study (after all four rounds of treatments).

In their November 2006 paper, the STAR*investigators also reported the outcomes from the maintenance phase, and this aspect of the paper had to baffle readers. These findings were of obvious critical interest to patients—with drug treatment, could they hope to stay well?—but the researchers spent little time discussing the results. In the abstract, they simply reported that "those who required more treatment steps had higher relapse rates during the naturalistic follow-up phase." Even in the text, the researchers spent only two paragraphs reporting the one-year outcomes, and they didn't give any specific numbers regarding relapse rates.

In order to discover a specific number, readers had to turn to Table 5 in the article. That table detailed the number of patients from each treatment step in the acute phase that had entered the follow-up phase as remitters, and the relapse rate for each group in the maintenance phase (levels 1 to 4). The rates ranged from 34 to 50 percent, and if readers got out their calculators they could discover that 568 of the 1,518 patients who entered the follow-up phase in remission had relapsed. This indicated that 950 patients—or 63 percent — had remained well.

There was one other graphic in the November report that seemingly showed relapse rates. Figure 3 provided specific numbers of nonrelapsed patients during the follow-up, but the numbers didn't appear to match the percentages listed in Table 5, and there was no explanation, either beneath the graph or in the text, that told of how the graphic should be read. It was evident that the STAR*D investigators had published some grouping of outcomes data in Figure 3 related to relapse rates, but it was nearly impossible to know what the numbers meant.

The NIMH Announces Results

In conjunction with the November article, the NIMH issued a press release announcing the results. "Over the course of all four levels, almost 70 percent of those who didn't withdraw from the study became symptom-free," the NIMH informed the public.[26] This was exciting news— "symptom free" told of patients who had been cured of an illness—and this also served as the bottom-line lesson from the study. "If the first treatment attempt fails, patients should not give up," said NIMH director Thomas Insel. "By remaining in treatment, and working closely with clinicians to tailor the most appropriate steps, many patients may find the best single or combination treatment that will enable them to become symptom-free."[27]

This became the finding remembered by the media. This was a study that had shown that antidepressants were quite effective in real-world patients. With the four treatment steps in STAR*D, *The New Yorker* reported in 2010, there was a "67 percent effectiveness rate for antidepressant medication, far better than the rate achieved by a placebo."[28]

The *New Yorker* is known for its vigorous fact-checking, and thus readers could expect that this report was accurate. The STAR*D investigators had promised that results would be rapidly disseminated and that they would guide future clinical care, and this was the key finding that was disseminated. Two-thirds of the patients had become "symptom free." The results from this $35 million NIMH study had been published and promoted in a way that preserved the market for the drugs.

STAR*D Deconstructed

After 2006, the STAR*D authors published dozens and dozens of articles, until the tally of such reports topped 100. But none of that literature upset the conclusions drawn in the three 2006 articles. And while anyone who closely read the three articles could sense that something was amiss, the precise trail of corruption detailed above may never have been known if it were not for the investigative work of psychologist Ed Pigott. He filed a freedom of information request seeking the protocol and other key documents, and once he had the protocol, which he has made available on the Internet, he could identify all the ways that the researchers had failed to follow it when reporting their results. Much of what Pigott reported in his two published articles has been incorporated into the above analysis.

In his two published articles and blog posts, Pigott also pointed out numerous errors of a more minor kind. "These errors are of many types, some quite significant and others more minor," he wrote. "But all of these errors—without exception—had the effect of making the effectiveness of the antidepressant drugs look better than they actually were, and together these errors led to published reports that totally misled readers about the actual results."[29]

Most important, it was Pigott who finally made sense of Figure 3 in the November 2006 article. That graphic, once deciphered, told of the number of remitted patients who had stayed well and in the trial throughout the 12 months. Of the 1,518 patients who had entered the follow-up in remission, only 108 had stayed well and in the trial throughout the 12 months. All of the others had either dropped out or relapsed back into moderate depression (or worse). Given that 4,041 patients had entered the study, this represented a documented stay-well rate of 2.7 percent at the end of one year.[30],*

Pigott and several collaborators published this finding in *Psychotherapy and Psychosomatics*. While the paper didn't attract the attention of any major media, *Medscape Medical News* did write about it, and the *Medscape* journalist turned to Maurizio Fava, one of the principal investigators in the STAR*D trial, for a comment. *Could this possibly be true?* "I think their analysis is reasonable and not incompatible with what we had reported," Fava said.[31]

In that way, Fava confirmed that Pigott had read the graphic correctly, and that his deconstruction of the entire study had merit. As for Pigott's paper not being "incompatible" with what the STAR*D investigators had reported, in a sense Fava was correct. The STAR*D investigators had

Table 7.3 Financial disclosures of STAR*D investigators (Number of ties to pharmaceutical companies)

Investigator	Academic Affiliation	Has Served as Advisor, Consultant or Speaker, or Received Research Funds	Tie to Forest, Maker of Study Drug
John Rush, M.D.	Southwestern Medical Center (U Texas)	10	Yes
Madhukar Trivedi, M.D.	Southwestern Medical Center (U Texas)	21	Yes
Stephen Wisniewski, Ph.D.	Univ. of Pittsburgh	1	
Andrew Nierenberg, M.D.	Massachusetts General Hospital	16	Yes
Jonathan Stewart, M.D.	Columbia Univ.	5	
Michael Thase, M.D.	Univ. of Pennsylvania	16	Yes
Philip Lavori, Ph.D	Stanford Univ.	6	Yes
Patrick McGrath, M.D.	Columbia Univ.	5	
Jerrold Rosenbaum, M.D.	Massachusetts General Hospital	26	Yes
Harold Sackeim, Ph.D.	Columbia Univ.	6	
David Kupfer, M.D.	Univ. of Pittsburgh	6	Yes
Maurizio Fava, M.D.	Massachusetts General Hospital	33	Yes

Source: Disclosure statement in A. John Rush, "Acute and longer-term outcomes in depressed outpatients requiring one or more treatment steps." *Am J Psychiatry* 163 (2006): 1905–17.

* Many of the 108 stay-well patients may have had initial HAM-D scores of less than 14, and thus weren't depressed enough to be eligible for the trial in the first place. Moreover, relapse in the maintenance phase of the study was defined as a HAM-D score of 14 or higher. Thus, it is possible that some of the "stay-well" patients were in fact worse off at the end of the study than at the start.

published Figure 3 in their November 2006 report. The data were there. But given that the graphic was so hard to decipher, it is fair to conclude that the STAR*D investigators had failed to make this bottom-line result *known*.

Economies of Influence

Although this was an NIMH-funded trial, industry influence was indirectly present during the trial. Rush and at least seven other STAR*D investigators had financial ties to Forest Laboratories, the manufacturer of Celexa.[32] The investigators' collective disclosure statement revealed hundreds of ties to pharmaceutical companies, with many investigators reporting that they had served as both consultants and speakers (Table 7.3). Yet, given that this was an NIMH-funded trial, STAR*D couldn't be blamed on the drug companies, and it could be argued that the "corruption" seen here far outstripped anything seen in a commercial trial of the SSRI antidepressants.

Rehabilitating SSRIs in Children: The TADS study

While the use of antidepressants in adults may have been questioned at times, the APA and academic psychiatry were nevertheless able to maintain a general societal belief in their merits. Adult use of antidepressants in the United States continued to steadily increase after the STAR*D study, and by 2008, 11 percent of American adults were taking an antidepressant on a daily basis.[33] However, after the FDA announced in 2004 that most pediatric trials of SSRIs had failed and that these drugs also doubled the risk of suicidal ideation in youth, it seemed that the pediatric market for SSRIs might collapse. Both the *British Medical Journal* and *Lancet* published articles concluding that prescribing these drugs to youth was contraindicated.[34,35] But then the results from the NIMH's Treatment for Adolescents With Depression Study (TADs) began appearing in the medical literature, and this study helped preserve the market for pediatric use of at least one SSRI, fluoxetine.

The Published Findings

In the TADS study, 439 youth aged 12 to 17 years old were randomized to placebo, fluoxetine (Prozac), cognitive behavior therapy (CBT), or a combination of CBT plus fluoxetine. At the end of 12 weeks, the response rate was highest for the combination group (71 percent), and lowest for the placebo group (35 percent).[36] While there were several academic critics who questioned aspects of this study—for instance, they noted that the only blinded comparison in the study was between fluoxetine and placebo, and in that instance, the drug had not provided a statistically significant advantage over placebo—this criticism was mostly ignored, and the TADS study was seen as providing convincing proof of the efficacy of fluoxetine in depressed adolescents.

However, the safety data—at least as initially reported—did provoke a note of concern. Nine of the 109 youth (8.3 percent) treated with fluoxetine alone suffered a "suicidal event" during the 12 weeks, which was more than double the rate in the placebo group (3.7 percent.) In 2006, the TADS investigators published a more comprehensive review of the safety data, and this time the reported numbers were slightly different, but even more concerning: ten of the fluoxetine-alone patients had experienced a suicidal event (9.2 percent), while there were only three such events in the placebo group (2.7 percent).[37] The researchers didn't conclude that this data showed that there was an increased suicide risk with fluoxetine, but the data spoke for itself: it was consistent with the FDA's black box warning that SSRIs increased the risk of suicide in youth.

In 2007, the researchers reported on the 36-week results for the three treatment groups. Although the placebo group was omitted from this report, there was still a group receiving CBT alone, and the suicide risk related to fluoxetine was pronounced: 16 of the 109 fluoxetine patients had suffered a suicidal event at the end of nine months (14.7 percent), compared to 7 of 111 (6.3 percent) in the CBT-alone group. But the researchers, in their report, didn't focus on this doubling of the suicide risk in the fluoxetine group. Instead, they noted that only 8.4 percent in the fluoxetine-plus-CBT group suffered a suicidal event, and thus, in the abstract, they drew this conclusion: "Adding CBT enhances the safety of the medication."[38]

The risk of *drug-induced* suicide seemed to be disappearing as a concern, at least in that abstract. The medication was safe, and adding CBT made it even safer. In March 2009, the TADs researchers published data that furthered this belief. After the first 12 weeks, the youth initially randomized to placebo could choose one of the three active treatments (fluoxetine, CBT, or a combination of the two), or opt to "not receive treatment" at all. And now the researchers put the focus on a new question: Had initial randomization to placebo *harmed* these youth? Did it increase their risk of suicide in this second part of the trial, from weeks 12 to 36? The researchers reported that twelve of the 112 patients (10.7 percent) initially randomized to placebo had a suicidal event during this second part of the trial, which was similar to the combined rate for the "active treatment groups (9.8 percent)."[39] Thus, the researchers concluded that initial treatment with placebo "does not increase harm-related events, including suicidality." The data also seemingly supported this conclusion: fluoxetine doesn't increase the risk of suicide either, since the rate of suicidal events in the placebo group was higher than the combined rate in the "active treatment groups."

Table 7.4   The TADS suicide data by initial randomized groups

Initial Randomization	At 12 Weeks		12 to 36 Weeks		Total	
	Suicidal Ideation	Suicidal Attempts	Suicidal Ideation	Suicidal Attempts	Suicidal Ideation	Suicidal Attempts
Non-Drug						
Placebo	5	0	1	6	6	6
CBT	4	1	0	2	4	3
Total Non-Drug	9	1	1	8	10	9
Fluoxetine						
Fluoxetine	9	3	1	3	10	6
Fluoxetine Plus CBT	3	2	3	1	6	3
Total Fluoxetine	12	5	4	4	16	9

In the TADS study, there were 439 youth aged 12–17 randomized into one of four groups: placebo (112); CBT (111); fluoxetine (109); and fluoxetine plus CBT (107). This chart shows suicidal events according to initial randomization, which was the data used by the investigators to report that fluoxetine didn't increase the risk of suicide. It appears that there are nine suicide attempts in both the non-drug and Prozac groups.
Source: B. Vitiello. "Suicidal events in the treatment for adolescents with depression." *J Clin Psychiatry* 70 (2009): 741–7.

In May 2009, the TADS researchers published a paper that focused specifically on suicidal events in the study (Table 7.4), and in this abstract, they put worry about fluoxetine-induced suicide to rest. There was "no difference in [suicidal] event timing for patients receiving medication versus those not on medication," the researchers wrote. There simply wasn't "evidence of medication-induced behavioral activation as a precursor" to a suicidal event. The TADS investigators had given fluoxetine a clean bill of health.[40]

Upon Further Review

As had been the case with the STAR*D, alert readers of the TADS articles might have sensed that something was not quite right in the way that the suicide data was being presented. However, it wasn't easy to precisely identify what was amiss. Then, in a 2012 blog, a Swedish psychiatrist, Göran Högberg, told readers where to find the true suicide data: there was a chart in that May 2009 paper that plotted the suicidal events in each of the four randomized groups according to when the events occurred, and whether the person was on an SSRI at the time of the event. The TADS authors didn't summarize the data in that chart, but it is easily done, and that summary tells of an astonishing result.

After the first 12 weeks, many in the placebo group had opted for treatment with fluoxetine. That was also true of some in the CBT-alone group. And, it turned out, it was *after* going on fluoxetine that these "placebo" youth and "CBT-alone" group became suicidal. Here is the real suicide data from the study:

- Seventeen of the 18 youth who attempted suicide during the 36 weeks were on fluoxetine at the time of their attempt. No patient on placebo during the 36-week trial had a suicide attempt. The only nondrug suicide attempt during the trial occurred in the CBT-alone group at week five.
- There were 26 other "suicidal events" in the study (preparation for suicidal behavior or suicidal ideation). Nineteen of the 26 events occurred in patients on fluoxetine. Three occurred in patients on placebo, and five in the CBT-alone group. (Thus, in total, 36 of 44 suicidal events occurred in youth on fluoxetine; Table 7.5).

The conclusion that could be drawn is obvious: This TADS study, when the data was summarized in appropriate fashion, corroborated the results from the randomized clinical trials of SSRIs for children and adolescents, which had shown that these drugs increased the risk of suicidal behavior in this age group. This study also added one very important

Table 7.5   The TADS suicide data based on drug exposure

	At 12 Weeks		12 to 36 Weeks		Total	
	Suicidal Ideation	Suicidal Attempts	Suicidal Ideation	Suicidal Attempts	Suicidal Ideation	Suicidal Attempts
Non-Drug						
Placebo	3	0	0	0	3	0
CBT	4	1	0	0	4	1
Total Non-Drug	7	1	0	0	7	1
Fluoxetine						
Randomized to Placebo	2	0	1	6	3	6
Randomized to CBT				2		2
Randomized to Fluoxetine	9	3	1	3	10	6
Randomized to Fluoxetine Plus CBT	3	2	3	1	6	3
Total on Drug	14	5	6	12	19	17

Many of the youth initially randomized to placebo or CBT went on fluoxetine during the 36 weeks, and only became suicidal after exposure to the drug. This chart groups suicide events according to drug exposure, which reveals that 17 of 18 suicidal attempts—and 82 percent of all suicidal events—occurred in youth on fluoxetine. But this bottom-line finding was not reported in the abstract or discussed in the text. It had to be dug out from the data presented in a table.

Source: B. Vitiello, "Suicidal events in the treatment for adolescents with depression." *J Clin Psychiatry* 70 (2009): 741–7.

detail to that finding: the increased risk appears to extend beyond suicidal thoughts to actual suicidal attempts. The result that needed to be known by the public was that 17 of 18 suicidal attempts had been in youth taking the antidepressant. Instead, the researchers told of how CBT enhanced the *safety* of the medication, and how an examination of the timing of suicidal events showed that the drugs didn't increase the risk of suicide. The TADS researchers did publish the relevant data—it was all there in that one chart—but they diverted the readers' attention elsewhere in their discussions, and ultimately they drew a conclusion belied by the data.

As was the case in the STAR★D trial, industry influence was present in this study. Five of the principal investigators in the TADS study had served on Eli Lilly's speaker's bureau; six more reported other types of financial ties to the company (honorariums, research support, and consulting services).[41] A number of the academic psychiatrists reported ties to numerous other manufacturers of SSRIs. However, once again, this was a NIMH-funded study, and thus the most direct "economy of influence" present was the investigators' guild interests. And in this study, the researchers had obscured the risk of drug-induced suicidality, which in some way, was similar to what GlaxoSmithKline and its thought leaders had done in their infamous study 329 of Paxil.

Markets Preserved

In each of the three studies discussed in this chapter, the results, if they had been presented clearly to the public, would have helped people better understand the risks and benefits of psychiatric drugs, and that understanding, in turn, may have led to a decrease in the use of these medications. The prescribing of stimulants for ADHD may have declined, and fewer adults may have chosen to take an antidepressant if they had known that so few people got well and stayed well over the longer term. The prescribing of SSRIs to children and adolescents had declined slightly after the FDA placed its black box warning on the drugs, and if the TADS suicide data had been accurately described to the media, then this decline might have continued and even accelerated—the risks with such treatment might have come to be seen by the public as outweighing any potential benefit. But in each case, the results were presented in a manner that preserved societal belief in the efficacy of these drugs.

In this book, we have been writing about two economies of influence. The pharmaceutical influence is well recognized by our society. There is much less societal attention paid to the guild influence in psychiatry, and how that interest may have a corrupting effect. The public likes to think of NIMH-funded studies as more "pure" than industry-funded ones. What the deconstruction of these three studies shows is that isn't necessarily so. Both economies of influence remain, with the guild interest paramount, and the three serve as examples of the resulting corruption.

CHAPTER EIGHT

The End Product: Clinical Practice Guidelines

It is difficult to think of any arena involving information about medications that does not have significant industry financial or marketing influences...Such widespread corporate interests may contribute to self-selecting academic oligarchies, narrowing the range of acceptable clinical and scientific information or inquiry.

—David Antonuccio, 2003[1]

The ultimate purpose of studies of medical treatments is to provide evidence that will lead to a "best use" of that intervention. The hope is that research can provide an in-depth understanding of a treatment's risks and benefits, both over the short and long term. The evidence base for any treatment should also provide insight into how a treatment may affect different patient subgroups. For instance, treatment outcomes may differ according to such variables as the severity of the patient's illness. Expert panels that review the research literature then develop "clinical practice guidelines" (CPGs) for use by the medical community, and in those documents, they can assess the merits of competing therapies. The CPGs are seen as the gold standard of "evidence-based medicine."

At first glance, it would seem—based on the problems with the research literature documented in the previous chapters—that the APA's clinical practice guideline committees could not hope to produce high-quality CPGs. Negative trials of psychiatric drugs have regularly gone unpublished; the research literature is filled with ghostwritten articles; and there are numerous examples of published articles that reported findings that were discordant with the actual study data. If the scientific literature is corrupted, how can a review of that literature lead to "evidence-based" practices? The answer to that question can be found in one of the fundamental principles of evidence-based medicine, which is that those who produce CPGs should review the research literature with a critical eye. They should assess the quality of studies, and discuss the limitations of the research literature and the ways that even meta-analyses may produce misleading findings. Thus, given that the problems with the research literature in psychiatry have become well known, particularly within the research community, the APA and its CPG panels, as they issued clinical practice guidelines, had an opportunity to restore a scientific integrity to their field. They could incorporate the deficiencies in the medical literature into their review process. They could reveal, in their discussion of their recommendations, that they had taken into account the negative studies, and that they had discounted results from industry-funded studies. They could also review the evidence from longer-term studies funded by the NIMH, such as the MTA study and the STAR*D study, which produced such dispiriting results. In this way, the CPG panels could exhibit a public willingness to think critically—and skeptically—about the profession's primary treatments.

In this chapter, we review the APA's development of clinical practice guidelines with that thought in mind: Is there evidence of a critical assessment of the merits of psychiatric drugs? Or is there, instead, evidence that the APA's clinical practice guidelines bear the imprint of the twin economies of influence that we have been writing about?

The Advent of Evidence-based Medicine

Although physicians have long sought to ground their practices in science, the specific concept of "evidence-based medicine" is of relatively recent origin. The history of this idea goes back to the early 1980s, when a group of clinical epidemiologists at McMaster University published a series of articles in the *Canadian Medical Journal* on the necessity for physicians to develop the skill to *critically* appraise research findings. In 1990, Gordon Guyatt, residency director of the internal medicine program at McMaster, who was one of the first to promote this notion of "evidence-based medicine," summarized this idea in a document prepared for incoming residents. He wrote:

> Residents are taught to develop an attitude of "enlightened skepticism" towards the application of diagnostic, therapeutic, and prognostic technologies in their day-to day management of patients. This approach...has been called "evidence-based medicine."...The goal is to be aware of the evidence on which one's practice is based, the soundness of the evidence, and the strength of inference the evidence permits.[2]

Practitioners in other medical specialties quickly embraced this fundamental principle: There was a need to critically appraise the quality of evidence produced by research studies in order to draw conclusions that could drive good clinical practice. As Kay Dickersin, director of the clinical trials center at Johns Hopkins School of Public Health, wrote, "tragedy can result from paying attention to poor quality evidence instead of good quality evidence."[3] By the mid-1990s, medical schools were regularly espousing the virtues of this approach, and as a result, a methodology for producing "evidence-based" guidelines was established. CPG panels were to conduct a systematic search for evidence, explicitly evaluate the quality of the evidence, and then make recommendations based on the best available evidence. In addition, a CPG review needed to provide an "assessment of the benefits and harms of alternative care options," the Institute of Medicine noted.[4]

Evidence-based medicine was heralded as ushering in a new era of care. In the past, the individual physician was seen as possessing the clinical expertise to decide what was best for patients. But this led to a great variation in how physicians practiced, which in turn often led to less than optimal outcomes. Evidence-based medicine would provide a systematic, objective method for guiding clinical practice. In addition, the older paradigm of care was a paternalistic one—the doctor knows best. Evidence-based medicine, with its careful assessment of different treatments, could provide a foundation for a patient-choice model, with the patients collaborating with physicians in reviewing options and choosing between competing therapies. In 2001, the *New York Times* touted evidence-based medicine as one of the best "ideas of the year."[5] A few years later, the *British Medical Journal* listed it as one of the 15 greatest medical milestones since 1840.[6]

The Subjectivity of CPGs

In the early 1990s, when the idea of evidence-based medicine was taking hold, there was a confidence within the medical establishment that "objective" clinical practice guidelines could be produced. There was a systematic process to be followed, and because the guidelines were based on a review of empirical data, they would not be vulnerable to the idiosyncrasies—or biases—of "expert opinion." It should not matter who the CPG panel members were: different panels could be expected to come to similar conclusions.

However, by the end of that decade, it had become evident that the composition of the panels did make a difference. CPG panels had to make many subjective decisions—what evidence should be considered, how the quality of trials should be judged, what outcomes truly mattered— and the composition of the panels could affect that decision-making. "All guideline committees begin with implicit biases and values, which affect the recommendations they make," noted Terrence Shaneyfelt, a physician from the University of Alabama School of Medicine, in a 2009 editorial in *JAMA*.[7]

The most obvious source of potential bias was a financial one. Although having a tie to industry should not be seen as evidence that a panel member is necessarily biased, it has been recognized that it can influence decision-making. And given that most CPG panels are formed of "experts" in their particular medical specialty, the members of the panels regularly have such ties. In a study of 44 CPGs for common adult diseases published between 1991 and 1999, researchers found that 87 percent of the CPG committee members had a connection to a pharmaceutical company, and on average, the CPG authors interacted with more than ten firms each. Fifty-nine percent of the CPG members had ties to the companies whose products were specifically considered in the guidelines.[8] In addition to that influence from pharmaceutical ties, there was also bias that arose from guild influences. The major funders and publishers of CPGs are medical societies and professional associations (such as the APA), and thus the panels for those groups may be affected by those professional interests. As Shaneyfelt wrote in his editorial, a CPG committee can "use guidelines to enlarge that objectively area of expertise in a competitive medical marketplace."

The committee members of a CPG, of course, may not be conscious of how such influences may affect their assessments of the evidence base. Those who are deemed experts in a particular discipline may have played a leading role in developing, testing and promoting new therapies, and thus are likely to be subject to a strong confirmation bias. This is the tendency for people to look for evidence that confirms their ideological or theoretical beliefs, and to ignore or discount evidence that belies or challenges those beliefs. This may occur even when scientists are certain they are objectively evaluating the evidence on a topic.

As researchers have studied such influences and looked for evidence of confirmation bias in CPGs, they have identified numerous examples of how a committee's makeup led to differences in recommendations. Most notably, they have compared CPGs issued by professional societies with a self-interest to those issued by a scientific group without a self-interest, and documented cases where the two groups produced CPGs with diametrically opposed recommendations. Those with a self-interest—and panel members with ties to pharmaceutical companies—may recommend a brand-name product over a generic; a scientific group without a self-interest may recommend the opposite. A professional society may recommend screening for a disease, while a scientific group without a self-interest may find that the benefit of screening does not outweigh the risks.[9] And so forth. As one researcher studying this problem noted, bias born of self-interest "almost always results in an overestimation of benefit [of a proposed intervention] and an underestimation of harm."[10] As a result, guidelines may end up favoring drug treatments over less expensive approaches—such as lifestyle changes—that have proven to be effective.

In a 2011 report, the Institute of Medicine (IOM) did not mince words about the scope of the problem: "Most guidelines used today suffer from shortcomings in development. Dubious trust in guidelines is the result of many factors, including failure to represent a variety of disciplines in guideline development groups, lack of transparency in how recommendations are derived and rated, and omission of a thorough external review process."[11] The fact that so many CPG panel members had ties to pharmaceutical companies threatened the integrity of the process as well. The IOM recommended that "whenever possible" authors of CPGs should have no such financial conflicts of interest, and that, at a minimum, those with such conflicts should comprise a minority of the CPG work group. The chair of any work group simply shouldn't have such a tie, the IOM stated.[12]

The APA's CPG Committees

The APA published its first practice guideline in 1991, and over the next two decades, it published at least 13 more. The APA produced CPGs for all of the major psychiatric disorders, such as schizophrenia, depression, bipolar disorder, and the anxiety disorders. These CPGs, the APA has stated, "provide evidence-based recommendations for the assessment and treatment of psychiatric disorders."[13] The guidelines are widely used and followed by psychiatrists and other physicians, such as family practitioners who prescribe psychiatric drugs, and thus can be said to set the standards that govern mental health care in the United States.

Psychiatry, as it developed such guidelines, faced challenges that made this task particularly difficult. In most other medical specialties, diagnoses are bounded by biological markers, and thus outcomes can be more easily quantified. Does a treatment reduce the size of a tumor? Lower blood pressure? Reduce cholesterol? Reduce or eliminate a virus? And so forth. However, in psychiatry, there are no biological markers that separate a patient with a "disease" from someone without it, which renders psychiatry more vulnerable to bias, since it relies more heavily on subjective judgments for making diagnoses and assessing symptoms. A clinician may see improvement in a psychiatric patient given a treatment when that same patient, on a self-rating scale, doesn't report feeling better. Psychiatry, in so many ways, is a discipline rooted in subjectivity, and yet, as the APA sought to develop CPGs, it was promising to produce recommendations that would be seen as objective, based on empirical data. As Irish psychiatrist Pat Bracken wrote in the *British Journal of Psychiatry*: "We [psychiatrists] will never have a biomedical science that is similar to hepatology or respiratory medicine, not because we are bad doctors, but because the issues we deal with are of a different nature."[14]

The other problem for psychiatry was that by the mid-1990s, nearly all of its academic leaders had ties to pharmaceutical companies. Such conflicts of interest weren't unique to psychiatry, but they were particularly pervasive in this field, and these financial ties—precisely because psychiatric diagnoses and outcomes are not bounded by biological markers—can influence psychiatrists in a particularly pronounced way. If any discipline had reason to establish CPG panels composed of a multidisciplinary group of researchers, without ties to pharmaceutical companies, it was psychiatry, but the APA's panels were regularly composed of academic psychiatrists with extensive ties to pharmaceutical companies.

In a 2009 review of the CPG panels for depression, bipolar disorder, and schizophrenia, researchers found that 18 of the 20 members had financial ties to the pharmaceutical companies that manufactured the medications recommended in the guidelines as first-line treatment (e.g., had equity in a company that made the medication, was a consultant or corporate board member, or received honoraria). Fourteen of the 20 worked as consultants to pharmaceutical companies. On the CPG panel for bipolar disorder, all of the members had ties to the companies whose drugs were identified as "first-line pharmacological treatment." The CPG for depression listed nine drugs as optimal medication, and all of the companies that manufactured one of those nine drugs had a financial tie to one or more of the panel members. The same was true for schizophrenia: the CPG listed 16 drugs that were commonly used, and every manufacturer of one of those drugs had a financial tie to one or more of the schizophrenia panel members. None of the three CPGs disclosed the financial conflicts of interest of the panel members.[15]

Since that study was published, the APA has updated its guideline on major depressive disorder, producing a new CPG in 2010. It did so during a period when the Institute of Medicine was raising concerns about the need to reduce financial conflicts of interest in CPG panels. There were six members of the APA group that produced the 2010 CPG, and every one reported a commercial tie to one or more of the companies that manufactured the antidepressants recommended in the CPG (collectively, the six members disclosed more than 120 such ties).[16,17] Four of the six, including the chair, Alan Gelenberg, served on speakers' bureaus or advisory boards of companies that sold antidepressants. Gelenberg had served as a consultant to 14 pharmaceutical companies, and been a member of the speakers' bureau for Eli Lilly, Pfizer, and GlaxoSmithKline, the manufacturers of best-selling SSRIs. He is also the long-time editor-in-chief of the *Journal of Clinical Psychiatry*, which as the *Carlat Report* has noted, is the "preeminent producer of pseudo-CME supplements" that are funded by industry.[18]

The 2010 CPG contains over 1000 references, 106 of which were papers that had been authored by one of the panel members. On average, each panel member's work was cited nearly 20 times in the guideline (range 0–58.) Most of these articles were research reports, reviews, or editorials involving a pharmaceutical.[19]

In the CPG, the six authors did disclose their financial ties to industry, providing a transparency that is often promoted as a way to eliminate bias from such conflicts of interest. The APA also set up a five-member "independent" panel to review the CPG for evidence of industry influence, and this panel gave the CPG a clean bill of health. Although the committee may have had extensive ties to industry, there was "no evidence" of bias in the CPG, the review panel concluded.[20] However, the review group did not provide a description of how it came to that conclusion, and while the CPG stated that the members of this review group had not had ties to industry in the previous three years, a search of public disclosures reveals that three of the five had—at some point—ties to pharmaceutical firms that manufactured antidepressants, with such ties including having served as consultants and on speakers' bureaus.[21]

The Skepticism Is Missing (and More)

As noted earlier, a CPG work group is supposed to conduct its evaluation of the research literature with an attitude of enlightened skepticism, which means it should assess the "soundness of the evidence, and the strength of inference the evidence permits." The committee should detail the critical thinking that led to its recommendations. The committee should evaluate alternative care options and describe the decision-making that led to its recommendations. While it may be difficult for any CPG to identify the "truth" about the relative merits of different therapies, a CPG should at least reveal a search for "truth."

With that understanding in mind, what follows is a quick review of the APA's *Practice Guideline for The Treatment of Patients with Major Depressive Disorder.*

Assessing the Efficacy of Antidepressants

The APA's committee members considered double-blind, randomized clinical trials to be the best evidence for determining the short-term efficacy of antidepressant drugs. They conducted a thorough search of all such studies that had been published since 2000, writing that this would update the reviews and meta-analyses of randomized clinical trials (RCTs) that had been conducted for previous CPG guidelines for depression. Based on this review, the committee concluded that "a large body of literature supports the superiority of SSRIs compared with placebo in the treatment of major depressive disorder," with the other types of antidepressants— tricyclics and serotonin-norepinephrine reuptake inhibitors (SNRIs)— said to provide a similar benefit over placebo.[22]

RCTs, of course, can vary greatly in study quality. However, the APA's work group, in its review of this evidence, did not fully address such issues as duration of study, study design, or the end-points used to assess efficacy. There was little discussion of the limitations of the RCTs for antidepressants, and particularly of this fact: Approximately 75 percent of clinical trial results published in major journals are industry funded, and when that is the case, it is 5.3 times more likely that the study will find a positive result for the drug than in studies that aren't commercially funded.[23,24]

The next problem with the RCT literature for antidepressants is that negative studies have gone unpublished. As Turner reported in 2008, 36 of 74 industry-funded studies of SSRIs and other new antidepressants failed to find a short-term benefit for the drug over placebo. Nearly all of the negative studies had either gone unpublished or been spun into positive ones.[25] The published literature reported 49 positive studies and only three failed trials. Turner's article had served as a wake-up call to promoters of "evidence-based medicine" precisely because it revealed that the published literature of RCTs of antidepressants was grossly misleading. However, the APA's CPG committee did not report Turner's findings, and it did not discuss, in any detail, its implications. Instead, the committee simply wrote that while "selective publication of positive studies could affect the apparent effectiveness of treatment, these factors do not appear specific to particular medications or medication classes."[26] If that sentence is carefully parsed, it's easy to see that it is directing the reader's attention away from the important implication of Turner's findings, which is that RCTs have often found that antidepressants do not beat placebo, and to a mostly irrelevant point, which is that his findings don't alter any assessment to be made of the *relative* merits of different antidepressants. The sentence ultimately functions as a "don't worry" comment to readers; there is nothing to be overly concerned about here.

Finally, the committee needed to address the evidence that had been reported by Irving Kirsch and Jay Fournier that SSRI antidepressants have not been shown to provide a clinically meaningful benefit, compared to placebo, for patients with mild to moderate depression. In 2008, Kirsch reviewed the data submitted to the FDA for four antidepressants (Prozac, Effexor, Serzone, and Paxil), and he found that drugs provided a clinically significant benefit "only for patients at the upper end of the very severely depressed category."[27] Jay Fournier at the University of Pennsylvania investigated this question of efficacy in a different way. As he noted, "there has been a paucity of investigations of the true effect of [antidepressants] in patients with less severe depression."[28] He and his colleagues searched the published literature for studies that reported on results in patients grouped by the severity of their symptoms, and also didn't use a placebo washout design. A placebo washout excludes those who initially respond well to placebo, a design that is common in industry-funded trials and is expected to suppress the placebo response, and thus make the drug look better by comparison. Fournier found only six studies in the literature that met his inclusion criteria, and in a meta-analysis of those studies, he found that "true drug benefits were nonexistent to negligible among depressed patients with mild, moderate and even severe baseline symptoms, whereas they were large for patients with very severe symptoms."[29] It was only in patients who were very ill that antidepressants provided a clear benefit.

Together, the Kirsch and Fournier studies presented compelling evidence that SSRIs and other new antidepressants increased the probability that a patient would improve more on drug than on placebo in only a small group of patients—those who were very severely depressed. Yet, the

APA's CPG committee did not accurately present this finding in its report, or even discuss it in a substantive way. Instead, in a one-sentence note, the committee wrote that "response rates in clinical trials typically range from 50 percent to 75 percent of patients, with some evidence suggesting greater efficacy relative to placebo in individuals with severe depressive symptoms as compared with those with mild to moderate symptoms."[30] The CPG committee was referring to Kirsch's and Fournier's work, but citing it as evidence that the drugs were highly effective for all patient groups. The only apparent question, based on this sentence, was whether they were even more effective in severely depressed patients than in those with milder problems.

In this way, the APA's CPG committee concluded that there was an "evidence base" for prescribing antidepressants to all depressed patients as a first-line therapy, regardless of the severity of their symptoms.

Assessing the Real-World Effectiveness of Antidepressants

Clinicians relying on a CPG may not appreciate that there is a difference between the "efficacy" and "effectiveness" of a medical treatment, with the latter measure being the more important one. A randomized clinical trial assesses the "efficacy" of a drug treatment over a short period of time, using an outcome measure, the World Health Organization (WHO) has stated, that "may or may or may not be clinically relevant." As was noted in the previous chapter, such trials also typically exclude patients with comorbid conditions, and thus the patients in RCTs may represent only a small fraction of the "real-world" patients that doctors see. Effectiveness, the WHO has written, measures the "likelihood and extent of desired clinically relevant effects" in the entire gamut of real-world patients, and over longer periods of time.[31]

Researchers in psychiatry understood that clinical trials did not provide evidence of effectiveness in real-world patients. The APA's CPG committee even addressed this in its guidelines, specifically discussing the difficulties "translating efficacy evidence to clinical practice."[32] The citation in that section leads to the study that John Rush conducted in 118 "real-world" patients. Rush's findings, of course, revealed that antidepressants were *not* effective in the real world. Even though he provided the best care possible, response and remission rates in real-world patients were "remarkably low."[33] This was the real-world result that readers of this guideline—that is, other psychiatrists and physicians—would want to know, but the APA committee did not discuss it. Instead, the CPG panel just discussed the limitations of RCTs in general terms, and thus Rush's data, which is so relevant to questions of drug effectiveness, didn't inform the evidence base the committee relied upon to make its recommendations.

The CPG committee had another study it could turn to for this purpose: the STAR*D trial. It too was conducted in "real-world" patients.

This study also failed to show that antidepressants were effective at the end of one year in such patients. The STAR*D investigators, at the onset of the trial, had even said that its "results should have substantial public health and scientific significance, since they are obtained in representative participant groups/settings, using clinical management tools that can easily be applied in daily practice."[34] However, even though the CPG committee cited the STAR*D study at least 13 times, it never mentioned the poor one-year outcomes. Instead, it relied on the STAR*D study to discuss drug-switching or drug-augmentation strategies for patients who hadn't responded well to a first antidepressant. The STAR*D trial, the CPG committee wrote, "provided evidence for continued efficacy of medication augmentation."[35]

Assessing Long-term Effectiveness

Both Rush's first study in real-world patients and the STAR*D study raised an obvious concern: Are antidepressants helpful over the long term? What makes this question of particular relevance is that any thorough review of the outcomes literature would reveal that, for nearly 40 years, at least a few researchers had publicly worried that antidepressants may *worsen* the long-term course of depression (in the aggregate).

This concern first arose in the early 1970s, when a handful of psychiatrists noticed that their patients, after taking an antidepressant, seemed to be relapsing more frequently than in the pre-antidepressant era. The drugs seemed to be causing a "chronification" of the disease, one psychiatrist wrote.[36] A Dutch physician then investigated this question and determined that it was so, concluding in a 1973 report that long-term use of antidepressant medication "exerts a paradoxical effect on the recurrent nature of the vital depression. In other words, this therapeutic approach was associated with an increase in recurrent rate."[37] Studies found that patients prescribed antidepressants were indeed relapsing frequently, so much so that the APA, in its textbook, began describing how depression "is a highly recurrent and pernicious disorder."[38] Next, the NIMH, in 1990, reported the results from an 18-month study that compared a tricyclic antidepressant (imipramine) to two forms of psychotherapy and placebo. Only 19 percent of the imipramine group were still in remission at the end of the period, which was the lowest for any of the four groups.[39]

Finally, in 1994, Giovanni Fava, editor of *Psychotherapy and Psychosomatics,* wrote that it was time for psychiatry to open "a debate on whether [antidepressant] treatment is more damaging" than helpful over the long-term.[40] Could it be, Fava wondered, that antidepressants induced a change in the brain that made the patient more biologically vulnerable to depression? After that first article, Fava repeatedly sounded this theme, marshaling evidence that "use of antidepressants may propel the illness to a more malignant and treatment unresponsive course" over the long term.[41]

While this was an alarming possibility, Harvard Medical School psychiatrist Ross Baldessarini, who has long been considered one of the preeminent researchers of psychiatric drugs in the world, wrote that, given the data that Fava had marshaled, it was one that deserved "open-minded and serious clinical and research consideration."[42]

A number of studies published since 1994 have provided more evidence that the drugs may have this adverse long-term effect. In a six-year NIMH-funded study of 547 people, researchers found that the "untreated individuals...had milder and shorter-lived illnesses," and that they were much less likely to end up "incapacitated" than those who took antidepressants.[43] Then there was Rush's study of real-world patients, and the STAR*D trial, both of which produced such terrible stay-well rates at the end of one year. As such data piled up, Fava regularly revisited this concern, and a handful of researchers even began positing a biological explanation for why antidepressants might be depressogenic agents over the long term. An SSRI, they noted, increased the levels of serotonin in the synaptic gap. But in response to this "perturbation" of neurotransmitter function, the brain, in an effort to maintain a "homeostatic equilibrium," dials down its serotonergic machinery. The density of receptors in the brain for serotonin decreases, such that the brain ends up with a serotonergic deficiency. Antidepressants, Fava wrote, may "worsen the progression of the disease in the long term by increasing the biochemical vulnerability to depression."[44]

However, the APA's CPG committee did not discuss any of this history. Instead, it recommended that antidepressants could be used as a "maintenance therapy." If this drug treatment failed, the APA recommended that the patient could be "given continuation ECT."[45] Any worry about the long-term negative effects of SSRIs is simply absent from the CPG, even though this is information that prescribing physicians and their patients would surely want to know.

A year after the APA published its depression guidelines, Rif El-Mallakh, an expert in mood disorders at the University of Louisville School of Medicine, investigated this question of the long-term effects of antidepressants. He conducted a review of the relevant outcomes literature, which was the same literature that the APA's committee had at its disposal, and he concluded that antidepressants may induce a chronic depressive state, which he dubbed "tardive dysphoria." El-Mallakh wrote:

A chronic and treatment-resistant depressive state is proposed to occur in individuals who are exposed to potent antagonists of serotonin reuptake pumps (i.e. SSRIs) for prolonged time periods. Due to the delay in the onset of this chronic depressive state, it is labeled tardive dysphoria. Tardive dysphoria manifests as a chronic dysphoric state that is initially transiently relieved by—but ultimately becomes unresponsive to—antidepressant medication. Serotonergic antidepressants may be of particular importance in the development of tardive dysphoria.[46]

The evidence base told of such a possibility, but the APA's CPG committee did not address this issue.

Comparing Alternative Treatments

The CPG committee did assess a number of "alternative therapies," including St. John's wort and exercise. As for St. John's wort, the committee noted that whereas studies in Europe had found the herbal remedy better than placebo for major depression, in the two largest controlled studies in the United States, it had failed this test.[47] One of the two was the NIMH-funded study led by Jonathan Davidson, which found that neither St. John's wort nor Zoloft was superior to placebo; however, the committee did not mention that Zoloft also failed to show efficacy in that trial.[48] As for exercise, the committee in its review did find evidence that it led to a "modest improvement" in mood disorders.[49] One of the studies that the committee cited compared exercise to Zoloft to Zoloft plus exercise, but the committee did not report the long-term outcomes from that study: at the end of 10 months, 70 percent of the exercise-alone patients were well, compared to less than 50 percent of the Zoloft-exposed groups.[50] At least in that study, exercise proved superior to drug and to drug plus exercise (as opposed to simply leading to a "modest" improvement).

Recommendations

The committee's recommendations are set forth in a summary, which undoubtedly is the part of the document read by busy practicing physicians. The committee recommended antidepressants as a first-line therapy for *all* patients with major depressive disorder, including those with "mild to moderate symptoms."[51] ECT was recommended as a "treatment of choice" for patients with moderate to severe symptoms who hadn't responded to antidepressants or psychotherapy, or for those patients who, regardless of severity of symptoms, "prefer" it.[52] Psychotherapies of various types were recommended for patients with mild to moderate symptoms.[53] For patients who show no or limited improvement after four weeks on an antidepressant, the committee recommended that other drugs be added—lithium, a thyroid hormone, or an atypical antipsychotic.[54] As for alternative therapies, St. John's wort was not recommended; the summary also did not mention exercise as a recommended treatment.[55] (In the body of the guidelines, the committee did write that "if a patient with mild depression wishes to try exercise alone for several weeks as a first intervention, there is little to argue against it."[56] In other words, this alternative therapy, while not exactly recommended, probably wouldn't hurt.) There is one

Table 8.1   CPG recommendations for initial treatment for mild depression

Intervention	US Guidelines	UK Guidelines	Netherlands Guidelines
Antidepressant	Yes	No	No
Psychotherapy	Yes	Yes	Yes
Antidepressant plus psychotherapy	Yes	No	No
Other somatic therapies	Yes	No	No
Lifestyle	No[a]	Yes	Yes
Low-intensity psychological[a]	No	Yes	Yes

[a] The APA's clinical practice guideline recommends that "if a patient with mild depression wishes to try exercise alone for several weeks as a first intervention, there is little to argue against it."
[b] For example, guided self-help or computerized cognitive-behavioural therapy.

Source: Derived from L. Cosgrove, "From caveat emptor to caveat venditor: time to stop the influence of money on practice guideline development." *J Eval Clin Prac* 20 (2014):809–12.

line in the recommendation section that somewhat endorses an alternative treatment: bright light therapy, the committee wrote, "might be used."[57] This was the committee's tepid endorsement even though, in a later discussion of the evidence, it concluded that "bright light therapy appears effective for seasonal affective disorder and nonseasonal major depressive disorder...Bright light therapy is a low-risk and low-cost option for treatment."

The APA's recommendations stand in sharp contrast to the recommendations of the National Institute of Clinical Excellence (NICE) in the United Kingdom. NICE is composed of a multidisciplinary group of experts, and given that research has shown that antidepressants do not provide a clinically meaningful benefit to patients with mild to moderate symptoms, the NICE guidelines recommend nonpharmacologic therapies as first-line treatments for that group of patients.[58] NICE also cautions against augmenting an antidepressant with an antipsychotic, noting that "patients whose antidepressant is augmented by an antipsychotic are much more likely to leave treatment early because of side effects."[59] National guidelines in the Netherlands also advise that when risk–benefit ratios for all treatments are assessed, nonpharmacological therapies should be the first-line treatment for mild-to-moderate depression (Table 8.1).[60]

As Shaneyfelt wrote in *JAMA*, the panel members of a CPG may have implicit biases that affect their decision-making and thus ultimately affect their recommendations. In the case of the APA's guideline for depression, the panel did not present the findings from Kirsch, Fournier, and Turner in an accurate manner. It did not present the findings from Rush's study of real-world patients; and it did not present the poor long-term results from the STAR*D study. Those are decisions that, whether consciously made or not, privileged guild interests. The panel's recommendation of antidepressants as a first-line intervention for mild to moderate depression did so as well. The APA's CPG for depression, while presented as an "evidence-based review," may also be understood as a document that protects and enhances the market interests of the profession.

ADHD Guidelines: A Second Case Study

The APA is not the only psychiatric institution in the United States that issues practice guidelines for the profession. In pediatric disorders, the American Academy of Child and Adolescent Psychiatry (AACAP) has taken a lead in issuing CPGs, and in a 2014 article, researchers at Harvard Medical School, led by Michael Murphy, compared the AACAP guidelines for ADHD to the NICE guidelines for this condition.[61] The two CPGs stood in even greater contrast than the CPGs from the two countries for adult depression.

The AACAP committee, which published its guidelines for ADHD in 2007, was led by Steven Pliszka, a professor of psychiatry at the University of Texas Health Science Center at San Antonio. The published guidelines do not disclose his possible conflicts of interest, or even list other members of the committee.[62] However, Pliszka has disclosed in other publications that he has served as a consultant and speaker for Shire Pharmaceuticals, which manufactures Adderall and Vyvanse, two popular ADHD medications.[63]

Pliszka and his committee recommended stimulants as a first-line treatment for children and adolescents of all ages. The guidelines also recommend behavioral therapy as an initial treatment if the child's symptoms are mild with minimal impairment, or if the diagnosis of ADHD is uncertain, or if the parents object to medication. If the child does not respond well to the stimulant, then the clinician is advised to consider "the use of medications not approved by the FDA for the treatment of ADHD." The recommended off-label medications included antidepressants and antihypertensive agents; this was a recommendation, Murphy noted in his article, that opened the "door to polypharmacy." Pliszka and his colleagues also concluded that studies had shown that Ritalin was "effective in preschoolers with ADHD," and thus was a recommended treatment, but with the thought that dosage "should be titrated more conservatively in preschoolers."

In contrast, the NICE guidelines recommend "group-based parent training and education programs" as the first-line treatment for school-age children diagnosed with ADHD. Drug treatment, NICE concluded, "is not indicated as the first-line treatment for all school-age children and young people with ADHD. It should be reserved for those with severe symptoms and impairment or for those with moderate levels of impairment who have refused nondrug interventions, or whose symptoms have not responded sufficiently to parent-training/education or group psychological treatment." The NICE committee also concluded that "there is no evidence" that drug treatment for ADHD is effective in preschool children, and thus is not recommended for that age group.[64]

Murphy and his colleagues did not try to assess whether one guideline was of higher quality than the other, or more consistent with the evidence. What is clear is that the AACAP guidelines protect a pharmaceutical paradigm of care, while the NICE guidelines challenge it.*

Expert Consensus Guidelines

Even before CPGs are produced, experts in a field may produce "consensus guidelines" that are expected to guide treatment. They serve as a first step in the production of recommended standards of care, even though they may simply represent the opinions of experts in a field, as opposed to arising from a systematic review of the evidence. As new psychiatric drugs were brought to market in the 1990s, academic psychiatry produced a number of such consensus reports. The authors of those reports regularly had ties to pharmaceutical companies, and their efforts to produce the consensus guidelines were often funded by "unrestricted" grants from the pharmaceutical companies that sold a drug for the disorder being studied. Finally, many of the consensus guidelines were published in journal supplements, with the *Journal of Clinical Psychiatry* a favorite repository for such reports, with these supplements paid for by pharmaceutical firms.

In 2010, David Rothman, a professor of social medicine at Columbia University Medical School, laid out in detail the making of one such consensus guideline and the economies of influence that were present.[65] He was hired as an expert witness by the state of Texas when it sued Johnson & Johnson for its illegal marketing of Risperdal, which provided Rothman with an opportunity to review internal Johnson & Johnson documents.

Shortly after Janssen, which is a division of Johnson & Johnson, received FDA approval to market Risperdal as a treatment for schizophrenia, it engaged three prominent academic psychiatrists to produce an expert consensus guideline for that disorder. The three psychiatrists were Allen Frances, chairman of the psychiatry department at Duke University; John Docherty, vice-chairman of the psychiatry department at Cornell University; and David Kahn, an associate clinical professor of psychiatry at Columbia University. Johnson & Johnson provided their three universities with an unrestricted grant of $450,000 to produce the expert consensus guidelines.[66]

---

* In 2014, Spanish investigators conducted an exhaustive review of the outcomes literature for stimulants, focusing on both their "short-term and long-term effectiveness," and concluded that "the result is disappointing and should lead to a modification of the CPGs to use of drugs as tools of last resort, in a small number of cases and for limited and short periods of time." Source: M. Valverde. "Outreach and limitations of the pharmacological treatment of Attention Deficit Disorder with Hyperactivity (ADHD) in children and adolescents and Clinical Practice Guidelines." *Rev Asoc Esp Neuropsiq* 34 (2014):37–74.

Seven months after signing the contract, Frances and his colleagues sent Janssen a draft of the "Tri-University Guidelines" they had developed. They asked the company "to please make comments and suggestions" on their work, and once they had this feedback, they published their guideline in a supplement issue of the *Journal of Clinical Psychiatry*. The publication was supported by "an unrestricted educational grant" from Janssen. The guideline recommended that Risperdal and other second-generation antipsychotics should be used as first-line treatments for schizophrenia, instead of the older agents like haloperidol.

Before the guideline was published, Janssen asked Frances and his colleagues for their assistance in marketing the guidelines, and getting them adopted by providers. In response, they formed Expert Knowledge Systems (EKS), informing Janssen that their company was ready to "move forward in a strategic partnership with Janssen" and help Janssen "influence state governments and providers" to adopt the recommendations in the guidelines. One of the ways this was then done was through Continuing Medical Education programs, where Frances, Docherty, and Kahn often appeared as speakers. Janssen paid EKS $427,659 for these services, money that Rothman stated went to the three academic psychiatrists, as opposed to their universities.

The Rothman report provided the public with an inside look at the financial conflicts of interest that were present as psychiatry produced its consensus guidelines. Experts with financial ties to pharmaceutical companies surveyed the opinions of other experts with ties to those same companies, and this led to the publication of a guideline in a journal supplement paid for by a pharmaceutical firm. The drug company that paid for all this with unrestricted grants could then use the guideline for marketing purposes, and it would pay the authors of the guidelines to serve on its speakers' bureaus, or as the experts at Continuing Medical Education programs it funded. There was a tight web of mutual interests present as psychiatry developed its expert consensus guidelines, and again and again, the experts recommended the new drugs as first-line treatments.

In his report, Rothman summarized the production of the schizophrenia "expert consensus guidelines" in this way: "From the start, the [Tri-University] project subverted scientific integrity, appearing to be a purely scientific venture when it was at its core, a marketing venture for Risperdal."

From A To Z

In this part II of the book, we have investigated, step by step, the conduct of American psychiatry over the past 35 years. We have investigated the APA's promotion of its DSM manual; the testing of new psychiatric drugs; the expansion and protection of that market; and finally the publication of expert consensus guidelines and CPGs that regularly recommended those newer drugs as first-line therapies. At every step, we have seen the integrity of the scientific process compromised by two economies of influence: psychiatry's "improper dependency" on the pharmaceutical industry, and psychiatry's own guild interests. Now we can turn our attention to the final part of the framework developed by the Safra Center for studying institutional corruption: the social injury caused by the corruption, the cognitive dissonance that makes it difficult for people within the institution to recognize the corruption, and prescriptions for reform. This is the ultimate purpose of our book: can investigating psychiatry through this institutional corruption framework illuminate possible solutions?

PART III

The Search for Solutions

CHAPTER NINE

A Society Harmed

Institutional corruption, which is usually built into the routines and practices of organizations, is usually more damaging to the institution and society than individual corruption, which in advanced societies typically consists of isolated acts of misconduct with effects limited in time and scope.

—Dennis F. Thompson, 2013[1]

The framework of institutional corruption provides a method for an investigation of an institution to proceed step by step. First, identify the institution's public mission, that is, its obligation to serve the public good. Next, identify the "economies of influence" that are present and may corrupt the institution. Then document the institution's "corrupt" behaviors. At that point, seek to understand the harm done to society by the corruption. This step, which can be quite difficult, provides a moment to pause and reflect, for the harm done may go beyond what is immediately visible, and instead reach deep into our collective lives.

From a big picture perspective, the harm done in this instance of institutional corruption arises from a very simple fact. Our society, over the past 35 years, has organized itself in response to a narrative told by American psychiatry that was, in so many of its details, misleading. Our understanding of the biology of mental disorders, our use of psychiatric drugs, our spending on psychiatric services, and even our social policies arise from a story that has been shaped by guild and pharmaceutical interests, as opposed to a narrative told by a medical profession that has shown an adherence to scientific principles and a commitment—at all times—to the best interests of patients. Those may be harsh words, but such is the comprehensive nature of the corruption documented in the previous chapters.

That corruption, in turn, translates into a long list of harms to society. There is harm done within a medical sphere (the diagnosing and treatment of psychiatric disorders), and also harm done within a larger arena, that of societal conceptions of what it means to be human. Organized psychiatry, with its DSM and its public narrative of how diagnoses represent brain diseases, has been presenting us, ultimately, with a new philosophy of being, and it is a philosophy that makes for a weaker, less resilient society.

Without Consent

The idea that patients, when confronted with a medical problem, should give their "informed consent" to any proposed treatment is grounded in the concept of personal autonomy, a cherished principle in the United States. The individual has the right to self-determination, and thus the doctor's duty is to provide information that will enable the patient to make an informed choice.

Although it might seem that this ethical principle would have been articulated long ago, the term "informed consent" is actually of fairly recent origin, first defined in the Nuremberg Code. The Nazis had conducted atrocious experiments on prisoners, without their consent, and the Nuremberg Code, which arose from the Nuremberg Trial of Nazi doctors who performed those experiments, identified "informed consent" as a fundamental principle in research involving human subjects. The volunteers needed to be well-informed about the risks of a study before they could give consent. In the 1950s and 1960s, this principle was extended to regular medical practice. A surgeon, for instance, needed to provide a patient with information about the risks of a proposed surgery, as well as its possible benefit. Then, in 1972, a landmark case in federal court set forth the *legal* obligation this imposed on physicians.

"The patient's right of self decision shapes the boundaries of the duty to reveal," the court ruled in *Canterbury v. Spence*. "That right can be effectively exercised only if the patient possesses enough information to enable an intelligent choice."[2] The case also set forth a standard for assessing whether this legal obligation had been met: "What would a reasonable patient want to know with respect to the proposed therapy and the dangers that may be inherently or potentially involved?"[3] A risk is considered material when a reasonable person, in what the physician knows or should know to be the patient's position, would be likely to attach a significance to the risk or cluster of risks in deciding whether or not to forego the proposed therapy.[4]

While it may be the individual physician who is supposed to obtain the "informed consent" of the patient, this legal standard clearly imposes an ethical duty—by proxy—on the medical specialty that provides individual physicians with the information that is supposed to be disclosed. The medical specialty must provide physicians with the best possible accounting of the risks and benefits of any proposed therapy, and, in its communications to the public, do the same. If the medical specialty doesn't do this, then the individual physician will not be capable of providing the patient with sufficient information "to enable an intelligent choice."

The diagnosis of a disease is obviously a first step in obtaining informed consent. What is the illness, or physical problem, that needs to be treated? If the symptoms do not lead to an easy diagnosis, that is understandable—that absence of knowledge helps inform the patient's decision making. If there is no known pathology causing the symptoms, that is fine too. Again, the absence of knowledge is important to the patient's consideration of therapeutic options. But if a patient is misled, and told he or she has a known pathology, when there isn't scientific reason to think that is so, then clearly the patient cannot then give "informed consent" to any subsequent treatment. Yet, that is precisely what has happened in psychiatry for decades. The American Psychiatric Association (APA) and the larger institution of psychiatry, in collaboration with the pharmaceutical industry, publicly promoted the "understanding" that psychiatric disorders were caused by a chemical imbalance in the brain, and that psychiatric drugs helped fix that imbalance, like "insulin for diabetes." American society as a whole came to understand that was true, and, as millions of Americans can attest, that was the message told by psychiatrists (and other prescribing physicians) to individual patients.

The informed consent standards, as set forth in *Canterbury v. Spence* and other court cases, provide a foil for assessing the depth of that ethical failure. Informed consent arises out of a duty to honor individual autonomy and the individual's right to self-determination. In this instance, psychiatry is informing the patient that he or she suffers from a brain illness, and that psychiatry has a drug to help fix that illness. The prescribed treatment will change how the brain works, and thus how the person experiences the world and responds emotionally to it. It is hard to imagine any decision more important to one's autonomy, and right to self-determination, than a decision of this type, and yet, because of the "chemical imbalance" story, psychiatric patients regularly made this decision within a context of misinformation. As might be expected, when people who have been told they suffer from a chemical imbalance subsequently discover that isn't so they may feel quite betrayed.★

That diagnosis may also limit further discussion of the merits of a proposed drug treatment. If a person suffers from a chemical imbalance, which a drug can then fix, like insulin for diabetes, then why wouldn't a person take such a drug? The "like insulin for diabetes" metaphor served as a summation of the risks and benefits of the proposed treatment: the introduction of insulin as a treatment for diabetes was one of the great medical advances of the first half of the twentieth century, and apparently psychiatric drugs were in that mold. Once that metaphor is presented to

★ We understand that there were psychiatrists who did not use this metaphor with their patients; we are writing here, as in the rest of the book, of common "institutional" practices in the field.

a psychiatric patient, the patient can be understood to be misinformed about the risks and benefits of the proposed treatment.

However, the individual patient, even if presented with that metaphor, may have an interest in being informed about the specific risks and benefits of a psychiatric drug that a physician is presenting as a proposed treatment. Society as a whole obviously has an interest in knowing this too. A medical specialty has a duty to provide information that will make this risk–benefit assessment possible, and yet, as was documented in the previous section of the book, psychiatry, while operating under twin economies of influence, has not produced a medical literature that can fill that need.

The psychiatric literature is flawed in numerous ways. Academic psychiatrists regularly lent their names to ghostwritten papers. Trial data may have been "mined" in order to report a positive finding for a drug that failed to show efficacy on a primary outcome measure. Conclusions announced in abstracts may be discordant with the data in the body of the article. Reviews of data submitted to the FDA reveal a different profile of risks and benefits for a particular drug than is found in the published articles. Poor outcomes for medicated patients in long-term studies have not been communicated to the public. And so on. All of this produces a medical literature that exaggerates the benefits of psychiatric medications and underestimates their risk, which makes it impossible for psychiatrists and other physicians to provide their patients with sufficient information to "make an intelligent choice." The published "evidence base" is tainted, and thus all of society ends up somewhat in the dark.

Finally, the clinical care guidelines developed by a medical specialty can be seen as part of this "informed consent" process. The developers of such guidelines are supposed to do a thorough search of the medical literature in order to set recommendations for standards of care. Both physicians and patients could then hope to rely on such guidelines as a proxy: experts in the field weighed the risks and benefits and drew conclusions for "best practices." Yet, the APA's guidelines often reflect an industry bias. For example, the recommendations promote antidepressants as a first-line treatment for mild to moderate depression, despite the fact that a critical analysis of the literature reveals that they do not provide a favorable risk–benefit ratio for this group of patients.

Add all this up, and it is fair to say that few of the millions of Americans who now take a psychiatric drug on a daily basis made an "informed choice" to do so. As an institution, American psychiatry did not create an environment for that to occur. Instead, most Americans have chosen to take a psychiatric medication within an environment riddled with misinformation (the chemical imbalance story), or a lack of information (the failure to disclose poor long-term results, for instance). Many individuals might be thankful for the drugs, while others may regret they ever took them, but those individual outcomes aren't the point here. The point is that our societal use of these medications has occurred without genuine informed consent.

Physicians Left in the Dark

As can be seen, informed consent standards impose an obligation on physicians to provide patients with the information about a proposed treatment necessary to make an "informed choice." Psychiatrists and other physicians who regularly prescribe psychiatric medications are thus relying on academic psychiatry and the APA, in its preparation of CPGs, to provide them with the information they need to practice in an ethical manner. When academic researchers lend their names to ghostwriting exercises, or fail to accurately report the results from their own trials, or develop industry-friendly CPGs, they are betraying the prescribing physicians too. We think of institutional corruption as harming the public while benefitting the members of the institution, but it is easy to see that practicing psychiatrists, who are surely motivated to see their patients do well, are harmed by this corruption as well. They have reason to feel betrayed by their own institution.

A Rising Burden of Mental Illness

As the APA promoted its disease model, it told a story of medical progress. The biology of mental disorders was being discovered, and, at the same time, the field was coming to better understand the true extent of mental illness in our society. For too long, these brain diseases—depression, various anxiety conditions, ADHD, juvenile bipolar disorder, and so forth—had gone underrecognized and undertreated (or simply hadn't been recognized at all). The APA's educational campaigns, which have been a constant for 35 years, informed the public of a pressing need for Americans suffering from such illnesses to get diagnosed and treated. The publicity campaigns did achieve their goal of ushering more people into treatment, and thus one might expect that this would lead to a lessening of the burden of mental illness in our society.

That has not been the case, however. Instead, over the past three decades, the burden of mental illness in our society has notably *worsened*. One measure of that can be found in disability numbers. In 1987, the year that Prozac came to market, there were 1.25 million adults, ages 18 to 66, receiving either a Social Security Supplementary Income (SSI) payment or a Social Security Disability Income (SSDI) payment because they were disabled by a mental illness. Twenty-five years later, the number of adults receiving a disability payment due to mental illness had topped 4.2 million.[5] Meanwhile, societal spending on psychiatric drugs rose from around $800 million in 1987 to more than $30 billion in 2012.[6]

The question that our society now needs to ask is whether this rise in disability is connected to psychiatry's behavior—its expansion of the boundaries of mental illness, and its campaigns to increase the use of psychiatric medications. Is there evidence that this is so?

First, it is notable that the sharp rise in disability is being driven by affective disorders, as opposed to psychotic disorders. The old mental hospitals, prior to the arrival of Thorazine in 1955, were filled mostly with psychotic patients. In 1955, there were 355,000 people in state and county mental hospitals with a psychiatric diagnosis, and of this population, only 50,937 had a diagnosis for an affective disorder.[7]* At that time, depression and manic–depressive illness were fairly uncommon disorders, and they were also understood to be episodic disorders, with patients expected to recover from an episode of illness and then often remain well for extended periods (if not indefinitely). But then, in the 1980s, the APA reconceptualized depression and anxiety as brain illnesses, which often required lifelong treatment, and the boundaries for diagnosing bipolar illness were expanded too. People so diagnosed began taking antidepressants, benzodiazepines, and mood stabilizers on a regular basis, and as that occurred, the number of people in our society disabled by such affective disorders steadily increased. In 2012, there were 2.1 million adults who received an SSI or SSDI payment due to a mood disorder.[8] That is more than twice the number of people receiving a disability payment due to schizophrenia.

The connection between increased treatment and disabilities is also seen in epidemiological studies. In 2005, Ronald Kessler, in an article on the National Comorbidity Survey, reported that the prevalence of anxiety, mood, and substance disorders among the adult population was 29.4 percent from 1990–1992, and that this prevalence rose only slightly—to 30.5 percent—in 2003. The number of people who fit diagnostic criteria for those disorders didn't change much during that 13-year period, evidence that social stresses were not stirring more anxiety and depression in the population. However, what did change during those 13 years was that the percentage of people with such disorders who got treatment jumped from 20.3 to 32.9 percent.[9] That would seemingly be a good thing, and yet as more people got treated during that 13-year period, the number of people receiving SSI or SSDI due to mental illness more than doubled, from 1.47 million people in 1990 to 3.25 million in 2003.[10]

Second, the APA, in conjunction with the pharmaceutical industry, has succeeded in exporting its DSM disease model to developed countries around the world. If this model of care is effective and helpful, then at least some of those countries should be seeing a moderation of disability rates. But if this is a model that increases the burden of mental illness in a society, then other countries should also be seeing rising disability rates due to psychiatric disorders. The latter is true. Here are just a few examples:

---

* There were another 210,000 people in the state and county mental hospitals with non-psychiatric disorders, such as alcoholism, Alzheimer's disease, and mental retardation.

- In Australia, there were 57,008 adults on government disability due to mental illness in 1990. That number rose to 241,335 in 2011, a fourfold increase.[11]
- In New Zealand, there were 21,972 adults, ages 18 to 64, on disability due to psychiatric conditions in 1998. Thirteen years later, that number had more than doubled, to 50,979.[12]
- In Iceland, with its stable population, the number of new cases of disability due to a psychiatric problem increased from 84 per 100,000 adults in 1992 to 217 per 100,000 adults in 2007.[13]
- In Denmark, there were 3,550 new disability awards due to psychiatric disorders in 1999; 11 years later, this number had jumped to 8,812.[14]
- In Sweden, about 25 percent of all new disability claims in 1999 were due to psychiatric disorders; by 2011, this percentage had risen to nearly 60 percent.[15]
- Finally, in Germany, the number of adults going on government disability because of a psychiatric disorder rose from 39,037 in 2000 to 70,946 in 2010.[16]

Third, it is easy to identify a possible iatrogenic mechanism at work in this rise in disability rates. The DSM, with its expansive criteria for diagnosing a psychiatric disorder, deems a large percentage of the population as mentally ill. This leads to the diagnosis of a great many people with difficulties—mild to moderate depression, anxiety, etc.—that, in the absence of drug treatment, could be expected to pass with time. But according to the DSM model, people diagnosed with a psychiatric illness suffer from a brain disease, which may require lifelong treatment. The person's future is reconceptualized, and while some may do fine on the medications long term, there is reason to worry, as was seen in chapter 7, that antidepressants and other psychiatric medications may worsen the long-term course of mental disorders. Thus, diagnosis plus treatment has the potential to turn an episodic problem into a chronic one, at least for some people, with symptoms becoming more debilitating over time. That is a process that could be expected to increase disability rates.

Perhaps many readers will still note this is correlative data. But clearly, at the very least, the rising disability numbers raise a concern for society, one worth mulling over as we try to assess the "social injury" arising from the "institutional corruption" that we are studying in this book.

Pathologizing Childhood

Prior to *DSM-III*, relatively few children and teenagers were diagnosed with mental disorders and medicated. While societal expectations of children, in the decades before 1980, may have varied from generation to generation, there was an enduring sense that, when judging children, one should remember that "children will be children," a sentiment that placed many difficult behaviors within the norm. The understanding was also that teenagers turned into young adults, and from that new perch of maturity, they could see the foibles and errant ways of their teenage years. The child grew up, or to put it into brain-development terms, the frontal lobes matured.

But the publication of *DSM-III*, and subsequent iterations of the DSM, created a different framework for parents and society to assess childhood behaviors and temperaments. The rebellious preschool child could now be diagnosed with "oppositional defiant disorder." Five- and six-year-old boys (and girls) who found grade school boring, and found sitting at a desk for hours on end a chore, could now be diagnosed with ADHD. The teenage girl who experienced emotional outbursts was a candidate for a diagnosis of depression, as was the sullen teenage boy. And if such mood episodes were interspersed with periods of elation, or wild passions, then bipolar disorder could be diagnosed. Angry youth, and particularly those in foster care, were also candidates for that diagnosis, or perhaps another one in the DSM that justified treatment with an antipsychotic (for behavioral control reasons).

Diagnosis, of course, leads to treatment, and thus the testing of medications for under-age-18 populations presented psychiatry with an obvious moral challenge. The field needed to assess whether diagnosing and medicating children would help them grow up and thrive. For a society headed down this path, there could be no question that was more important to answer. Was this, in fact, a story of medical progress, of psychiatric illnesses that had previously gone undiagnosed but were now being recognized and successfully treated? However, as was documented in previous chapters, this testing was done by a medical specialty laboring under two powerful economies of influence, pharma money and guild interests, and the result was that this basic scientific inquiry, so vitally important to our society, was corrupted.

The MTA study by the NIMH of the long-term effects of stimulants found that the drugs did not help children grow up and thrive. While stimulants might provide a short-term benefit, the drugs did not provide a long-term benefit, and, if anything, the medicated children fared worse at the three- and six-year follow-ups. One of the primary investigators, psychologist William Pelham, articulated the moral duty for psychiatry at that point:

> We had thought that children medicated longer would have better outcomes. That didn't happen to be the case. There were no beneficial effects, none. In the short term, medication will help the child behave better, in the long run it won't. And that information should be made very clear to parents.[17]

But the APA, in its communications to the public, never made that finding known. Instead, it touted the 14-month results from the MTA study as evidence that ADHD drugs provide a long-term benefit. Guild interests prevailed over its moral duty.

The pediatric testing of SSRIs is now understood, as *Lancet* wrote, to be a particularly shameful episode in psychiatry's recent history, with the pediatric researchers betraying the "trust that patients place in their physicians."[18] The published results regularly exaggerated the benefits of SSRIs in children and teenagers, and underreported possible harms.

There are many risks associated with the prescribing of stimulants and SSRIs to children that haven't been publicized. Perhaps the most problematic risk is that these drugs can induce mood instability that leads to a diagnosis of bipolar disorder. In this way, the fidgety child or depressed teenager may be turned into a lifelong mental patient. Stimulants, by their very nature, set up a mood cycle: the drugs have arousal effects, which are typically followed by dysphoric effects as the drug leaves the body. Meanwhile, in a review of SSRI trials, researchers found that 8 percent of youth treated with the drug experienced a manic or hypomanic episode, versus 0.2 percent of the placebo group.[19] Studies of youth diagnosed with bipolar disorder have regularly borne witness to this iatrogenic pathway, with the "illness" triggered by exposure to a stimulant or SSRI. For instance, 84 percent of the children treated for bipolar illness at the Luci Bini Mood Disorders clinic in New York City between 1998 and 2000 had been previously prescribed psychiatric drugs; upon initial diagnosis, before exposure to a psychiatric drug, "fewer than 10 percent" had bipolar symptoms.[20] As *Time* magazine reported in 2002, most children with bipolar illness are diagnosed with a different diagnosis first, with "ADHD the likeliest first call."[21]

A thorough review of the risks and benefits of psychiatric drugs in youth is beyond the scope of this book. The point here is that in the post *DSM-III* era, the diagnosis and treatment of youth with psychiatric drugs has proceeded in an environment where the APA and psychiatric profession regularly protected the image of the drugs, with the public left largely uninformed about possible harms from this practice. And what is now quite easy to document is that raising children in this new DSM environment, which combines diagnosis with drug treatment, is proving injurious, on the whole, to our society.

The data alarms, alerting us to worsening mental health in children and teenagers, have been sounding now for 20 years. To wit:

- From 1996 to 2007, the number of total days that children and adolescents were hospitalized for psychiatric diagnoses rose from 1.96 million days to 3.64 million.[22]
- In 2001, US surgeon general David Satcher warned that the psychiatric difficulties of children and teenagers constituted a "health crisis."[23]
- In 2002, *Psychology Today* reported in an article titled *Crisis on Campus* that an increasing number of college students, arriving on campus with an antidepressant were experiencing "first episodes of mania."[24]

- In 2007, researchers reported that the number of children discharged from hospitals with a bipolar diagnosis rose fivefold between 1996 and 2004.[25]
- In 2008, the US Governmental Accountability Office (GAO) reported that one in every 15 young adults, 18 to 26 years old, was now "seriously mentally ill." These youth are "functionally impaired," the GAO said.[26]

All of these reports told of how, in this new DSM-ordered environment, an increasing number of children were faring poorly. Diagnosis plus treatment may all too often lead to an increase in psychiatric problems over the long term, and a decrease in the resilience that may gradually accrue during childhood. Government SSI statistics provide more possible evidence of the harm done. In 1987, there were 16,200 children who received an SSI payment because they were disabled by a mental disorder (this excluded mental retardation). Such children comprised only 5.5 percent of the 293,000 children on the disability rolls. But then the diagnosing of children and youth steadily increased, with those children regularly medicated, and as this occurred, the number of youth receiving an SSI payment because of a mental disorder steadily climbed, year after year. In 2011, there were 728,008 youth under age 18 who received an SSI payment because of a mental disorder. This group now comprised 58 percent of the 1.3 million children on the SSI rolls (up from 5.5 percent in 1987).[27]

Those data reveal a new career path that has opened up in American society in the *DSM-III* era. A high percentage of the children and youth who receive an SSI payment for a mental disorder will then, at age 18, move onto the adult disability rolls, as they will be understood to be suffering from a chronic illness, which manifested early in their lives.

An Impoverished Philosophy of "Being"

The previous list of harms relate to psychiatry's impact on society in the medical realm. Institutional corruption led to the practice of medicine in a manner that violated informed consent standards, and it also took a toll on our society as measured by a worsening of mental health outcomes, among both adults and children. But the institution of psychiatry also has had a profound impact as the creator of a modern *philosophy*, which affects our understanding of what it means to be "normal," our internal narratives, our societal response to social injustice, and even our democratic values.

In every society, stories are told that provide an understanding of "human nature," and if there is one enduring theme of such stories, it is that humans are extremely emotional creatures. The Old Testament tells of humans often beset by greed, envy, lust, love, and—at the most extreme—homicidal impulses. In that context, the essential struggle for a person is within one's self. The emotions that produce sinful behavior should be kept at bay, and in this way a person may hope to behave in ways that are pleasing to God. The Greek epics depicted similar struggles, and if we leap-frog forward in literary history we quickly land on Shakespeare, whose characters are regularly gripped by fierce emotions, whether it be the pangs of great longing in Romeo and Juliet, or the murderous rages of Macbeth. Novels provide us with a similarly rich tapestry, and in all of this literature, from Greek times forward, we see variations on the same theme: humans struggle with their minds and emotions, which are rarely neatly ordered.

Freud's conception of the mind fits that theme. His brilliance was to place literary dramatizations of the human character into a psychological framework, with the different elements of the mind—the id, ego, and superego—regularly at war (and with so much of this struggle occurring outside the purview of consciousness). His psychological depictions of the mind reflect the physiology of the human brain, with its ancient reptilian center, overlaid by a mammalian cerebral cortex, and finally the exaggerated frontal lobes, with that part of the brain regularly given the task of creating a narrative for the "I" that makes some sense, regardless of how accurate that narrative may actually be.

The beauty of such literature, and why ultimately it can be so comforting to an individual, is that it provides an expansive view of human nature. We all struggle with our emotions and our minds, and rare is the human being who hasn't played the fool, or had his emotions spiral out of control. This is an understanding that nurtures empathy, both for others and for one's own self. It also provides us with a sense of what to expect from life. Happiness may visit now and then, but count your blessings when it does. Anxiety, grief, rage, jealousy, melancholy—expect such emotions to visit too.

The APA, with its DSM, provides society with a very different vision of human "normalcy," and what we can expect from our minds. Normalcy—as defined by the diagnostic constructs that tell of "illness"—is a very constricted space. Extreme emotions, painful emotions, and difficult behaviors all become "symptoms" of diseases. The mind, according to *DSM-III*, *-IV* and *-5*, is not supposed to be a chaotic place. Even a very common feeling of distress, anxiety, may be seen as "abnormal." In this way, the DSM provides society with a remarkably impoverished philosophy of being.

The writer Sam Kriss brilliantly illuminated this fact by writing a review of *DSM-5* as though it were a novel.[28] There is, he notes, a sense of "loneliness that saturates this work," the book written by a narrator who lacks an ability "to appreciate other people."

This is a story without any of the elements that are traditionally held to constitute a setting or a plot. A few characters make an appearance, but they are nameless, spectral shapes, ones that wander in and out of view as the story progresses, briefly embodying their various illnesses before vanishing as quickly as they came—figures comparable to the cacophony of voices in *The Waste Land* or the anonymously universal figures of Jose Saramago's *Blindness*. A sufferer of major depression and of hypochondriasis might eventually be revealed to be the same person, but for the most part the boundaries between diagnoses keep the characters apart from one another...The idea emerges that every person's illness is somehow their own fault, that it comes from nowhere but themselves: their genes, their addictions, and their inherent human insufficiency. We enter a strange shadow-world where for someone to engage in prostitution isn't the result of intersecting environmental factors (gender relations, economic class, family and social relationships) but a symptom of "conduct disorder," along with "lying, truancy, [and] running away." A mad person is like a faulty machine. The pseudo-objective gaze only sees what they do, rather than what they think or how they feel. A person who shits on the kitchen floor because it gives them erotic pleasure and a person who shits on the kitchen floor to ward off the demons living in the cupboard are both shunted into the diagnostic category of encopresis. It's not just that their thought-process don't matter, it's as if they don't exist. The human being is a web of flesh spun over a void.

Kriss titled his review "The Book of Lamentations." Ultimately, he concluded, the DSM presents readers with a description of normalcy that fills one with "horror":

If there is a normality here, it's a state of near-catatonia. *DSM-5* seems to have no definition of happiness other than the absence of suffering. The normal individual in this book is tranquilized and bovine-eyed, mutely accepting everything in a sometimes painful world without ever feeling much in the way of anything about it. The vast absurd excesses of passion that form the raw matter of art, literature, love, and humanity are too distressing; it's easier to stop being human altogether, to simply plod on as a heaped collection of diagnoses with a body vaguely attached.

This is the book that now provides our society with a framework for judging others, and ourselves. Emotions and behaviors that, in literary works from the past, would have been presented as quite normal are instead depicted as symptoms of "mental illness." The mischievous child—hello Tom Sawyer—is now recast as a child suffering from ADHD, who needs to be treated with a stimulant. The moody teenager, who once would have been seen as going through the "storm and stress" of adolescence, may now be diagnosed as depressed, and treated with an antidepressant.

The adult who grieves too long over the death of a spouse or a parent now has "major depressive order" (perhaps Shakespeare today would recommend a dose of Cymbalta for Hamlet). And so on. Try imagining what Zora Neale Hurston would be diagnosed with today, given how she described her life in *Dust Trucks on a Road*: "I have been in Sorrow's kitchen and licked out all the pots. Then I have stood on the peaky mountains wrapped in rainbows, with a harp and sword in my hands."[29] A physician conversant with the APA's diagnostic manual would have known just the remedy: A daily dose of Lamictal would have smoothed out Zora Neale Hurston's mood swings just fine.

In short, the DSM—as a book of philosophy for our modern society— squeezes the poetry out of life. It tells us to rein ourselves in, and to avoid feeling too deeply. And here is the tie to institutional corruption: The APA informed us of chemical imbalances and known brain illnesses and validated disorders, which are the building blocks for the philosophy it has created, and yet those building blocks have yet to be found in nature.

Ask Yourself: What's Wrong with You?

The APA and advocacy organizations, through their many educational campaigns, encourage the populace to be on guard for signs of "mental illness." In this way, psychiatry encourages the individual to adopt an internal narrative of a very particular sort. The "I" is being primed to ask, what is wrong with me? Am I anxious? Am I depressed? Am I distressed? Why can't I focus? Do I eat too much? Why am I angry so often? Why can't I be happy? The "I" is encouraged to be always on the alert for signs of *abnormality*, with parents taking on this vigilance for their children, even when they are of preschool age. Furthermore, this constant vigilance for signs of distress easily morphs into a larger life narrative. Wasn't I often unhappy as a child? Wasn't I regularly angry? The DSM then informs a person that if such emotions have been regularly present, these are symptoms of a "brain illness."

The solution to such distress, modern psychiatry tells the patient, lies outside the self, rather than inside. If one is suffering from a brain disease, with distressing emotions fixed by abnormal brain chemistry, what can one do other than take a pill? In this way, the APA's disease model encourages an internal narration that robs one of a sense of self-agency, and of resiliency. The person is not encouraged to examine the context of his or her life. What has happened? What might be changed? Or, even more radical: Are there times when suffering is to be embraced?

Erin's story tells of the harm that can result from adopting this type of internal narrative, and the benefit that can come from doing the opposite and examining the context of one's distress.[30] In her case, it was a psychologist who helped her see herself in a new light.

Today, at 31, Erin is engaging, keenly intelligent, and displays a confidence that, coupled with a genuine humility, draws people to her. She graduated magna cum laude from a good university, worked and saved money for six years, got married, and is now applying to Ph.D. programs in psychology. To look at her life now, it seems she is one of those people for whom things always came easily—school, friends, relationships. But, in fact, she had a difficult childhood. Her parents divorced when she was 11, the same year her mother was diagnosed with lymphoma, and soon she was having difficulty with her schoolwork and her friends. She struggled with anxiety, irritability, anger, and depression. She was told she had ADHD. She resisted that diagnosis, and only took the stimulant medication for a short time (which she found didn't help). During her senior year of high school, her mother's cancer worsened. Two very difficult years followed, with Erin prescribed medications for ADHD, depression, and anxiety, but the drugs never seemed to "fix" anything, or make her feel better. Then, in the fall of her junior year at college, her mother died. Now Erin struggled in every aspect of her life, her anxiety and depression worsening. Today, she recalls, with a self-deprecating laugh, she can't even begin to count "how many cell phones" she broke in fits of rage. Years after her mother died she remained depressed, and finally she gave in. There was something "wrong" with her. Maybe not ADHD, but something else. She went to see a therapist, constantly asking him to diagnose her, and "tell me what is wrong with me." She grew increasingly frustrated when he refused to do so. He wasn't giving her what the DSM said he should, and finally she pressed him further. At last, he responded: "Do you really want me to tell you what you have?" Relief flooded over her. "Oh thank God," she thought. "I finally got him to tell me, now I'll know what's wrong with me." The clinician then gave his diagnosis: "I think you have a broken heart."

Erin's internal narrative changed with that one sentence. As she puts it, telling this story 11 years later, "everything shifted." Instead of thinking that something was *wrong* with her, she realized that something had *happened* to her. Erin now viewed her past from the perspective of what she had been through, rather than what DSM disorder she *had* (i.e., what brain disease), and once she did that, a different future opened up for her. To have a broken heart, when you have suffered the losses she had, was to be normal.

A Barrier to Social Justice

As Kriss observed in his article, the DSM locates problems within the biology of the individual. This, in turn, is an understanding that encourages society to ignore social factors that may create mental difficulties for so many. If depression, anxiety, and other psychiatric symptoms arise from neurotransmitters that are out of whack, then society—much like the individual patient—is not prompted to consider the social context for such distress.

Perhaps the best example of this is how foster children are regularly treated today. Such children can be understood to have drawn the short straw in the lottery of life, born into a dysfunctional family situation, where abuse of many kinds may be rampant. Yet, such children today are often diagnosed with a mental disorder. Society is no longer prompted to ask what *happened* to the child; instead, a diagnosis is made to designate what is *wrong* with the child. The child is informed that he or she is broken and "mentally ill." Medication is offered as the solution, and this is so even though the medication may squelch the child's emotions and capacity to interact with the world. More disadvantages are heaped on the child, and all the while, society sees no need to create a nurturing environment for the child, one that would encourage the child to develop a new internal narrative of how he or she is capable of loving and being loved.

In this way, the disease model encourages society to turn a blind eye to the social injustices that may cause mental distress to so many. If anxiety, depression, or aggressive behaviors are due to a chemical imbalance, or a brain illness of some kind, then questions about lousy schools, poor wages, poverty, racism, and crime-filled neighborhoods can be dismissed. If a school doesn't set aside time for gym, art, or music, and the child gets antsy sitting in a chair all day, then that is the child's fault, and not the fault of a society that would create such a child-unfriendly school day. The DSM, all too often, breeds a societal hard-heartiness toward those who are suffering because of reasons related to social injustices, and mitigates the impulse for creating a more just society.

Not All Are Created Equal

The Declaration of Independence proudly declares that all men are created equal. The great struggle in American history has been to develop a society that, in its laws and behaviors, honors that sentiment (and incorporates women into that definition). That struggle requires the populace to avoid drawing lines that separate one group of people from another, or in any way designating certain people as "less than" others.

American society has often failed that test with regard to its treatment of the "mentally ill." The worst time may have come in the first half of the twentieth century, when eugenic attitudes took hold in America. Eugenicists portrayed the mentally ill as genetically unfit, and thus a danger to society. In order to keep the mentally ill from passing on their defective genes, eugenicists argued that they needed to be placed in institutions and kept there, at least through their breeding years, and that they should be forcibly sterilized, if necessary. These sentiments were translated into policy. People admitted to mental hospitals in the first half of the twentieth century often remained institutionalized for decades, and the US Supreme Court decided that forced sterilization of the mentally ill was constitutional. In addition, it was during that period, when the mentally ill were seen as having no worth, that lobotomy was popularized in asylum medicine (the 1940s). A surgery that quieted them, even if it removed the very part of the brain that distinguished humans from apes, could then be seen as worthwhile.

Today, we like to think that society has an "enlightened" attitude toward the mentally ill. The APA, the NIMH, and patient-advocacy groups conduct antistigma campaigns, which often carry the message that psychiatric disorders are brain disorders, and thus not anyone's fault. The person has a biological illness, which likely has a genetic component. However, in 2010, researchers reported that such campaigns, if anything, lead to an increase in stigma, as they found that people who believe "neurobiological conceptions of mental illness" are more prejudiced against the mentally ill (and thus, for example, are less willing to live next door to someone diagnosed with schizophrenia).[31] The problem, of course, is that the disease model, which claims to draw a boundary line between the "normal" and the "abnormal," fosters the idea of an "us" and a "them."

That is particularly true when it comes to societal feelings toward those diagnosed with schizophrenia or other psychotic disorders. With the disease model, there is little need to ask about the life stories of people so diagnosed, and see if they tell of trauma or of other life events that might drive one mad (which would engender empathy). These are people with a severe brain illness. And regardless of whether that is so, the conception of such people as the "other" leads society to fear them, and that in turn leads to the passage of laws that "us" would never tolerate. For instance, in recent years, numerous states have passed assisted outpatient commitment laws, which require people to take antipsychotics even when they are living in the community. These laws raise obvious issues related to civil rights, which is certainly how many people who have been forcibly treated see the issue, but because of our societal conceptions about schizophrenia, we pass such laws without discussing this civil rights aspect, and without providing a societal forum for listening to the voices of those who might be affected by such legislation. The DSM encourages us to think that people so diagnosed are "defective." Thus, as we pass outpatient commitment laws, we turn a deaf ear to their complaints about how the drugs may emotionally numb them, and leave them feeling like zombies.

Regardless of whether such medication is helpful, the passage of such laws occurs within a framework that is ultimately undemocratic. It tells of a society that is convinced that it has the right to curb the self-autonomy of a large group of people, who have violated no law, without any close examination of whether such "treatment" is helpful for them over the long term. There are many studies that in fact question that assumption. In the best long-term study ever conducted of schizophrenia outcomes, Martin Harrow reported that those patients who took themselves off antipsychotics had dramatically better long-term outcomes: they were eight times more likely to be in recovery at the end of 15 years, had better cognitive functioning, were less anxious and much less likely to still be suffering from psychotic symptoms.[32] Research has now shown that antipsychotics may shrink the brain, and that this shrinkage is associated with a worsening of negative symptoms and functional impairment.[33] But that information has not entered the public debate, and for that last fact, we can blame institutional corruption. If the APA publicized the results of the Harrow study, or the brain shrinkage studies, then society might have reason to question the morality of legislation that forces many people to take a medication that they feel robs them of a sense of being fully alive.

A Society Transformed

Such is the social injury exacted by institutional corruption within psychiatry. Psychiatry has been practicing in a manner that violates informed consent principles; the burden of mental illness in our society has dramatically increased in the past thirty years; the DSM provides our society with an impoverished philosophy of being; the APA's disease model may encourage individuals to adopt a debilitating internal narrative; and finally, this model may blind our society to social injustice and encourage antidemocratic sentiments. This is a story of institutional corruption that has exacted an extraordinary toll.

CHAPTER TEN

Putting Psychiatry on the Couch

It's difficult to get a man to understand something if his salary depends upon his not understanding it.

—Sinclair Lewis, 1935[1]

As a method of inquiry, a study of institutional corruption starts with the presumption that the individuals within the institution are "good people" who want to behave in an ethical manner, consonant with societal expectations and their own self-image as ethical beings. However, "economies of influence" encourage unethical or problematic behaviors throughout the institution, and eventually those behaviors become normative. At that point, you have a case of "good apples" working within a "bad barrel." Moreover, the fact that problematic behaviors have become normative may lead to an institutional blindness. Those within the institution lose the capacity to see themselves—or their institution—from an outsider's perspective. They will remain convinced that their behavior is ethical, confident that conflicts of interest have not altered their behavior, even while outsiders see their behavior as quite compromised, or even ethically outrageous.

That disparity in perspectives, between those within the institution and those on the outside, presents an obvious hurdle to any substantive public discussion about "reforming" the corrupted institution. If those within the institution don't see an ethical problem, then they cannot be expected to initiate reform, or even to welcome suggestions that such reform is needed. At the same time, a study of institutional corruption is expected to detail the social injury from that corruption, and thus to present society with a compelling rationale for action. What can be done to remedy this problem? As a first step toward "solutions," we need to understand the reasons that "institutional blindness" develops, and whether it is ever possible for a medical institution to see itself as corrupt.

The Science of Implicit Biases

Although physicians and clinical researchers may claim that they are unaffected by conflicts of interests, certain that their devotion to good science will enable them to remain "objective," there is abundant research showing that conflicts of interest, with great regularity, alter their behavior. A physician's prescribing patterns may change, and as for clinical researchers, their design of trials, their interpretation of results, their wording of abstracts, and their public discussion of study results may all be altered by such conflicts. However, the researchers may not be conscious of how the conflicts have led to a bias in their behavior, for the simple reason that bias, studies have shown, operates on unconscious parts of the mind.

While the idea that much of our mental activity is hidden to our conscious minds is not new, Harvard psychologist Mahzarin Banaji and her colleagues have recently demonstrated, with great clarity, how we may harbor "implicit biases" that are unknown to us and yet affect us in profound ways.[2] Of particular note, she developed an "implicit association test" that revealed a distressing truth about the legacy of racism in our country. In the test, people were asked to press keys to associate adjectives with black and white faces flashed on a computer screen, and with regularity they pushed keys more quickly when asked to associate a "good quality" with a white face, and a "bad quality" with a black face. Even people of color often harbored such "implicit attitudes," which revealed an internal bias that would have seemed foreign to many. Banaji's study also showed how such "implicit attitudes" translated into behavior: a slowing of one's physical ability to press the very key that would reveal us to be free of racial preferences. In this test, we want to press the "black is good" key as quickly as the "white is good" key, but because of unconscious "implicit attitudes," we may not be able to do so.

Banaji's work demonstrated the unconscious nature of "biases" that affect our behavior. Now imagine the effect of a financial conflict of interest upon behavior. The person is motivated, at an unconscious level, to behave in a way that protects the financial benefit (or other self-benefit), even if such behavior contradicts one's conscious code of ethical behavior. Research by neuroscientists illuminates why this is so. Decision making involves not just cognitive areas of the brain but emotional areas too. Imaging studies have shown that there is an integration of cognitive processes with emotion-processing areas of the brain such as the hippocampus and amygdala. The emotion-processing areas influence this decision making based on memories of previous experiences. Did a physician previously receive a payment to serve as an advisor to a pharmaceutical company? Was that physician treated in a particularly pleasurable way—perhaps a first-class air ticket, with car and driver waiting at the airport, and lodging in a first-class hotel? With such memories activated in the unconscious mind, the physician may then unknowingly opt to act in a manner that will keep such benefits flowing.

As is well known, industry-funded research regularly produces results that are more favorable to industry than independently funded research. The public is likely to assume that the investigators consciously decided, in some way or another, to produce a favorable result. But research on implicit bias suggests otherwise. For example, in one study, researchers evaluated the liver tissue of rats exposed to the drug dioxin, and those who were funded by industry identified fewer slides as cancerous.[3] But— and this is the important point—there was no evidence that the industry-funded researchers made a conscious decision to interpret the slides in a manner that was favorable to industry. Instead, it seems, the funding biased, at an unconscious level, their decisions about what they *saw* in the slides. The two sets of researchers, those with and without industry funding, literally saw the slides in different ways.

There are, of course, other reasons that industry-funded drug trials may so regularly favor the sponsor's drug. The company, if it controls the study, may take deliberate steps to produce a favorable result. What we are focused on here is the mindset of researchers who view themselves as independent scientists, and strive to act in that way. A conflict of interest will still create an internal bias that can affect the investigator's behavior and thought processes. This research also reveals why transparency—the disclosure of conflicts of interest—will not eliminate bias. The conflict remains, and that will affect the emotion-processing areas of the brain as the investigator conducts research, or performs other tasks. Disclosure will not alter that unconscious process.

Simon Young, coeditor-in-chief of *Journal of Psychiatry and Neuroscience*, summed up the problem in this way: "The idea that scientists are objective seekers of truth is a pleasing fiction, but counterproductive in so far as it can lessen vigilance against bias."[4]

Protecting the Self

Although the individual investigator may not recognize the bias, outsiders will, and that sets up a potential conflict for the investigator. Is it possible that he or she has acted in an ethically questionable manner? Cognitive dissonance theory explains what may happen next within the investigator's mind.

Cognitive dissonance theory grew out of research intent on understanding what people do when they are confronted with information that creates conflicted psychological states. Although Leon Festinger's original cognitive dissonance theory from 1957 has been revised multiple times, the basic premise remains that individuals experience psychological distress when their behavior is at odds with their ethical beliefs, or when they are trying to hold incompatible thoughts.[5] Individuals experiencing cognitive dissonance have a desire to reduce their feelings of discomfort by attempting to reconcile their conflicting beliefs and behaviors, or their incompatible thoughts, especially if the dissonance is esteem-related (e.g., is related to how one sees oneself professionally).

"Festinger's theory is about how people strive to make sense out of contradictory ideas and lead lives that are, at least in their own minds, consistent and meaningful," write Carol Tavris and Elliot Aronson, two social psychologists who have studied cognitive dissonance. "Because most people have a reasonably positive self-concept, believing themselves to be competent, moral, smart and kind, their efforts at reducing dissonance will be designed to preserve their positive self-images—even when that perpetuates behavior that is incompetent, unethical, foolish, or cruel."[6]

Much as implicit biases operate on an unconscious level, the resolution of such dissonant states also occurs outside the sphere of the conscious mind. In functional magnetic resonance imaging (fMRI) experiments of brain activity during dissonant settings, Tavris and Aronson note, "the reasoning areas of the brain virtually shut down." Then, when the dissonant state is resolved (i.e., the unconscious mind has presented a narrative to the conscious mind that preserves one's self-image or sense of being right), the "emotion circuits of the brain" light up. The mind is now *emotionally* pleased with itself. "These mechanisms provide a neurological basis for the observation that, once our minds are made up, it may be physiologically difficult to change them," Tavris and Aronson write.

Dissonant states may be resolved in numerous ways. We are more likely to notice and remember information that confirms our beliefs, and ignore or discount information that threatens our beliefs. If we are forced to confront information that threatens our sense of self, or threatens our beliefs, we may simply reject the information (it's just wrong), or metaphorically kill the messenger (he's biased, or has evil motives). If one's behavior verges on the unethical, and the person understands that this is the case, he or she may find reasons to justify it. "Everyone does it," or "I was just following orders," are common internal explanations.[7]

Numerous studies reveal that we are unaware of our own cognitive dissonance. In one study, researchers showed pictures of two women to the participants and asked them to choose the one they found more attractive. Then the researchers would either give them the photograph of the woman they preferred, or, using sleight of hand, hand them the photograph of the woman they had found less attractive. In either case, the participants were then asked to explain *why* they had found the woman in the photo they were now holding more attractive. Remarkably, 75 percent of the participants given the wrong photograph didn't notice the switch, and instead found reasons to explain their choice of this woman as the more attractive one. They might say she was "radiant," or notice that she, unlike the woman in the other photo, was wearing "earrings," which they liked. Their unconscious minds had constructed a narrative, on the fly, that presented their conscious selves with a coherent sense of self. They had seen this particular woman as more attractive from the start.[8]

We, of course, can easily recognize such cognitive dissonance in *others*. We see the self-justification, the selective filtering of information, the marshaling of new "facts," and the attribution of bias to others. However, researchers have found that it is difficult to see such machinations at work in our own minds, precisely because we need to see ourselves as rational. The mind rebels at the thought it constructs a subjective, and even untrustworthy, view of reality.

"In a sense, dissonance theory is a theory of blind spots—of how and why people unintentionally blind themselves so they fail to notice vital events and information that might throw them into dissonance, making them question their behavior or their convictions," Tavris and Aronson write. "Our blind spots allow us to justify our own perceptions and beliefs as being accurate and unbiased. If others disagree with us, we have two ways of reducing dissonance: we can conclude that we are wrong, or that they aren't seeing clearly. You know the preferred alternative."[9]

Finally, it is those with a strong sense of the self as moral and proper who may be most prone to suffering from cognitive dissonant states. The thief who robs a house does not suffer a dissonant state because he already knows he is a thief—that is who he is. But the good person who behaves in an unethical way is almost certain to suffer the pangs of a dissonant state, at least until it is resolved. "The nonconscious mechanism of self-justification is not the same thing as lying or making excuses to others to save face or save a job," Tavris and Aronson explain. "It is more powerful and more dangerous than the explicit lie, because it blinds us from even becoming aware that we are wrong about a belief or that we did something foolish, unethical, or cruel. Dissonance theory therefore predicts that it's not only bad people who do bad things. More often, the greater problem comes from good people who do bad things or smart people who cling to foolish beliefs, precisely to preserve their belief that they are good, smart people."[10]

Cognitive Dissonance in Doctors

Physicians, of course, are trained to see themselves as altruistic, guided in their actions by a desire to serve their patients' best interests. They spent years becoming doctors, and that considerable effort and sacrifice increases their self-image of being "moral" and "good." For the most part, society sees doctors in this light too. Thus, doctors have a keen need to protect that self-image, which may require a good dose of cognitive dissonance when their behaviors are influenced by conflicts of interest. Numerous studies have documented the presence of such cognitive dissonance in physicians, with much of this research focusing on the influence of pharmaceutical money on their behavior.

In a 2006 survey of obstetricians and gynecologists, Maria Morgan and colleagues found that the majority thought it was ethical to accept free drug samples (92 percent), a free informational lunch (77 percent), or a well-paid consultancy (53 percent). The physicians reasoned that the free sample would be helpful to patients in financial need (or provide added convenience), and only a third thought that their prescribing habits would be influenced by the free samples. However, they did worry about their peers; they were more likely to conclude that the "average doctor's prescribing would be influenced by acceptance of the items than their own."[11]

Similarly, in a survey of residents at a university-based program, Michael Steinman, at the University of California San Francisco, found that 61 percent of the young doctors believed that their prescribing patterns would not be influenced by gifts, yet thought only 16 percent of "other physicians" would be immune to such freebies. Gifts were a problem for others, but not for them. Moreover, with this self-image in mind, the majority of the residents found it "appropriate" to accept free meals, reprinted articles, pens, textbooks, and even to go on a free "social outing." The residents, Steinman concluded, "believe they are not influenced" by gifts from industry.[12] Indeed, another study found that the more gifts a doctor took, the greater his or her sense of being unaffected by them.[13]

The experts in a field, including key opinion leaders, may be particularly certain of their "objectivity" in spite of their financial ties to industry. Niteesh Choudhry, from the University of Toronto, surveyed 192 authors of 44 clinical practice guidelines endorsed by North American and European societies on common adult diseases, and found that 87 percent had ties to a drug company. On average, they had associations with more than ten companies. Nearly two-thirds of the authors (64 percent) served as speakers for drug companies, and 59 percent had relationships with the companies whose drugs were considered in the guidelines they wrote. Yet, only 7 percent of the authors thought that their financial ties to pharmaceutical companies "influenced" their recommendations, and a slightly higher number—19 percent—thought their coauthors were so influenced. In other words, more than 80 percent of the experts were confident that the financial relationships of the members of their group with pharmaceutical companies, including their own ties, did not influence the clinical practice guidelines they produced.[14]

Columbia University researcher Susan Chimonas found that physicians relied on various forms of denial and rationalization to manage their cognitive dissonance. "They avoided thinking about the conflict of interest, they disagreed that industry relationships affected physician behavior, they denied responsibility for the problem, they enumerated techniques for remaining impartial, and they reasoned that meetings with detailers were educational and benefited patients," she wrote.[15] Although the physicians' methods for resolving the conflict may have varied, typically the end thought was the same: It was okay for them to accept gifts because they would remain objective, even though others might be biased by such conflicts.

In addition, researchers have found that when physicians take gifts or are paid by pharmaceutical companies to serve as speakers or advisors, they may feel "entitled" to such rewards because of how much they sacrificed to become a doctor. In one study of this type, family medicine and pediatric residents were asked about their willingness to accept gifts, and if they were explicitly reminded of this sacrifice before answering the gift question, the percentage that thought it was fine to do so rose from 22 to 60 percent.[16]

The self-image of physicians as ethical professionals, noted former American Psychiatric Association president Paul Appelbaum, is so strong that for many even "the suggestion that they may be influenced by contact with the pharmaceutical or device industries is infuriating." As physicians take money and gifts from pharmaceutical companies, they need to see themselves as objective, acting in the best interests of their patients, and it is an affront to suggest otherwise. "For social psychologists who study the difficulties that people have recognizing how other parties influence their behavior, physicians' failure to appreciate the impact of relationships with industry merely makes physicians like everyone else," Appelbaum wrote.[17]

The Hidden Mind of Psychiatry

Research on implicit biases and cognitive dissonance explains why conflicts of interest may bias research, with the researcher being unaware of this effect and eager to defend his or her integrity as an "objective" scientist. But this book is interested in the corruption of an institution, which is to say the collective mind-set that developed within psychiatry. The premise here is that the twin economies of influence, pharmaceutical money and guild interests, created a deep-seated bias in that collective mind. And what research into implicit biases and cognitive dissonance reveals is that it may be nearly impossible for psychiatry to see its errant ways. If psychiatry is presented with evidence of corruption at the institutional level, or evidence that its therapies may not be particularly helpful, the field, in terms of its collective mind-set, will experience the same cognitive dissonance that an individual may experience when confronting a conflicted psychological state. The dissonance needs to be resolved, and psychiatry's leaders will employ the same unconscious strategies that individuals do when faced with dissonant situations: they will construct a collective narrative that protects the profession's financial interests as a guild and maintains the profession's self-image.

Protecting the Guild

In 1980, after psychiatry adopted a "medical model" for classifying mental disorders, the field was left with three main "products": research, the classification of mental disorders, and the prescribing of psychiatric drugs. As such, the field has an evident economic need to maintain societal belief in the integrity of its research, the validity of its diagnoses, and the merits of psychiatric drugs.

All medical disciplines have an interest in maintaining a belief in their therapies, but this is particularly true of psychiatry. In the United States, psychiatry is in competition for patients with psychologists, social workers, counselors, and other therapists who provide psychological services to people struggling with psychiatric issues. If psychiatric drugs are not seen as helpful, the field cannot hope to thrive in this competition. Patients will turn to the other providers of care. In addition, the prescribing of drugs has become central to what a psychiatrist does. Many psychiatrists today call themselves psychopharmacologists, and psychopharmacologists with ineffective drugs would not have much of a business. As Allan Detsky noted, such guild interests are a "form of bias [that] comes from the way you make your living."[18]

Given this guild interest, it is easy to identify research results that would provoke cognitive dissonance within the field. If a prominent study found a class of drugs to be ineffective, or, for instance, that unmedicated patients did better over the long term, then the APA and leaders in the field would need to find a way to dismiss the study in a manner that would protect societal belief in the medications. In a very real way, the APA and leaders in academic psychiatry must respond in this way. How could the field possibly admit to itself and to society that its main products might be doing more harm than good? A medical specialty can only abandon a therapy when it has a replacement in the offing, as otherwise it might as well sign its own "going out of business" certificate.

Psychiatry's Self-Image

In a 2010 study of physicians with "high but fragile self-esteem," University of Chicago investigators found that these individuals, when confronted with an "ego threat," were more likely than normal to respond in a manner that bolstered their self-image. Those with fragile self-esteem reported "greater self-perceived invulnerability to conflicts of interest, relative to their peers," the Chicago researchers reported. In other words, the ego threat provoked a greater sense of cognitive dissonance in those with "fragile self-esteem," an equation that has particular relevance to psychiatry, given its collective self-image.[19]

As is well known, psychiatry has long struggled with an inferiority complex in relation to other medical special specialties. Psychiatry had its roots in the asylum, and that beginning—and the fact that so many asylum patients remained unwell—helped nurture the notion that psychiatry was a medical backwater. That thought was an essential aspect of psychiatry's "crisis" in the 1970s. Insurance companies questioned the merits of its therapies, as did leaders in Congress. The antipsychiatry movement characterized psychiatrists as agents of social control, rather than as doctors tending to real illnesses. Then, when David Rosenhan published his 1973 article *On Being Sane in Insane Places,* it suddenly seemed that much of the scientific community was now laughing at psychiatry. "There is a terrible sense of shame among psychiatrists, always wanting to show that our diagnoses are as good as the scientific ones used in real medicine," wrote one psychiatrist, shortly after the APA voted homosexuality out of its diagnostic manual.[20]

Spitzer and his colleagues initiated their effort to remake the DSM because of a scientific impulse—they wanted to develop a manual that would be reliable. But there were also guild interests at work, with the APA, by adopting a disease model, eager to present itself to the public as a medical specialty that was part of mainstream medicine. *DSM-III,* said Gerald Klerman, a former director of the NIMH, "represents a fateful point in the history of the American psychiatric profession...its use represents a reaffirmation on the part of American psychiatry to its medical identity and its commitment to scientific medicine."[21] But this public pose was also one that psychiatry was presenting to itself: the field could now look in the mirror and see doctors dressed in white coats.

The self-image that psychiatry has cultivated ever since can be clearly seen in a sampling of talks by the leaders of the APA at their annual meetings over the past 30 years. These are speeches when psychiatry is, in essence, talking to itself, and such moments are vital to creating a "group" mind-set.

John Talbott, APA president-elect, 1984

Let me reemphasize that our relationship to the rest of medicine and organized medicine is at a critical phase. Through the hard work of thousands of our members at national, state, and local levels, we are once again being seen as 'real doctors' who are interested not only in their own specialty and subspecialties but in the problems, challenges, and actions of all of our medical colleagues. Thus, our continuing reintegration into medicine must proceed vigorously.[22]

Carol Nadelson, APA president-elect, 1985

Psychiatry has moved from a backwater to the forefront as a medical specialty, largely because of the research explosion, principally in the neurosciences.[23]

Robert Pasnau, APA president, 1987

During my lifetime, I have seen a change in the way that psychiatrists and psychiatric patients have been viewed. In my internship class of 16, none of the four of us who were planning careers in psychiatry dared to inform the selection committee of our interest for fear that we would not get our coveted positions. Many of us suffered demeaning jokes from our fellow residents in other medical specialties. I shall never forget the professor of surgery whose rounds I attended while still a resident. Upon learning that I was a psychiatric resident, he glowered at me and muttered, "What a hell of a waste of a medical education." While such outrageous stereotyping still occurs, there has been a profound change in the degree and valence of such attitudes.[24]

Paul Fink, APA president-elect 1988

I maintain that psychiatric stigma is best handled not by battering our heads against stone walls, not by insisting that people stop hating us, deriding us, undercutting us, and competing with us, but rather by establishing that we are at the forefront of the modern medical paradigm.[25]

Herbert Pardes, APA president, 1990

Psychiatry in 1990 is at the height of its powers. We have had a spectacular decade...Psychiatry works, psychiatry is respected for it, and we can hold our heads high.[26]

Joseph English, APA president, 1993

We have seen the growth of a strong and vibrant psychiatric profession whose new knowledge and treatments bring miraculous relief to mentally ill patients once without hope.[27]

Herbert Sacks, response to presidential address, 1997

Future historians of medicine will be astonished by two decades of revolutionary scientific advances in psychiatry, illuminated by new findings in the neurosciences, psychopharmacology, and dynamic treatment modalities.[28]

Steven Sharfstein, APA president, 2006

As physicians we search for the truth based on science, producing replicable results through research. The Scientologists protesting outside this meeting in Toronto represent the opposite of the search for truth. They are joined in a general movement against science by such groups as the intelligent design advocates, abstinence-only fanatics, global warming deniers, antivaccination lobbies, gay bashers, and stem cell research rejecters. There is a conflict going on in America today, sometimes called a "culture war," but very much a conflict between those who search for truth based on science to improve the human condition and those who are blinded by dogma and denial.[29]

Nada Stotland, presidential address, 2009

We must learn to celebrate ourselves. Our skills are unique and precious: medical and psychological knowledge, communication *and* compassion—all in the service of healing some of the most devastating diseases that afflict humankind. We are brain doctors *and* doctors of emotions and cognition. We chose to pursue medical *and* mental health training. We dissected cadavers. We attended operations, births and deaths...Every psychiatrist is a hero.[30]

Alan Schatzberg, presidential address, 2010

Psychiatry is a wonderful specialty that unlike others must face concerted and, at times, coordinated efforts of an antipsychiatry movement...Psychiatrists should take pride in their often heroic efforts to take care of those with mental illness. Not only are we dedicated physicians, but we lobby hard on the behalf of the disadvantaged.[31]

Carol Bernstein, president-elect, 2010

As psychiatrists, we have been stigmatized by our colleagues...how many of you were told that you were "too smart" or "too normal" to be a psychiatrist? How many were advised by your medical school deans to go into surgery, medicine, or orthopedics rather than psychiatry?[32]

Jeffrey Lieberman, APA president, 2014

We have been waiting, many of us our whole lives, for the chance to change the way the world thinks of psychiatry and the way we think of ourselves as psychiatrists. Let's use the momentum we have to plunge ahead into the next year with our confidence brimming, our energy renewed, and our sights set high...this is our opportunity to change the practice and perception of psychiatry for the better and as never before. Last year, standing on the stage in San Francisco, I told you that "our time has come." Today, I say to you that our future is now![33]

Those speech excerpts reveal how psychiatry strove to see itself. Psychiatrists are real doctors, the field's research has produced extraordinary advances, its treatments are highly effective, and the entire field is motivated—really in a heroic fashion—to serve its patients. At the same time, it feels disrespected. A psychoanalyst, listening to such talks year after year, might also have noticed a hint of institutional narcissism, which is typically related—as the University of Chicago researchers noted—to "high but fragile self-esteem." As such, when psychiatry is confronted with information that threatens its self-image, it might be expected that it will provoke a profound case of cognitive dissonance. Or, as the University of Chicago investigators gently concluded, "concerned about self-image, physicians in this situation may be insufficiently receptive to new information and instead attempt to justify initial opinions."[34]

Cognitive Dissonance on Display

Over the past 35 years, psychiatry has regularly been confronted with information and findings that threatened its self-image (and its guild interests). The field has come under a great deal of criticism for its close ties to the pharmaceutical industry. Psychiatry, in fact, was seen by many as having been "bought" by industry. Even more threatening, study after study questioned the merits of its drug therapies. The MTA study found that the stimulants did not provide a long-term benefit on any domain of functioning for children diagnosed with ADHD. Kirsch and Fournier, through meta-analyses of clinical trial data, reported that SSRIs, which the field had promoted as remarkably effective, did not provide any meaningful benefit to patients with mild to moderate depression. The FDA announced that most pediatric trials of SSRIs had failed and that they doubled the risk of suicidal ideation. The STAR*D produced extremely dispiriting one-year results. The new atypicals, which the field had touted as breakthrough medications in the 1990s, were subsequently found to be no better than the old medications. Harrow reported that unmedicated schizophrenia patients had much better long-term outcomes. And so on. Finally, even the chemical imbalance story began to publicly collapse, leaving psychiatry with the challenge of explaining why that story had been told to the public in the first place.

All of these findings and reports shared one common feature: They were discordant with the story that psychiatry had been telling to the public for the past 35 years, and to itself. Cognitive dissonance theory can help explain psychiatry's responses to these dissonant moments.

Bias? What Bias?

In 2007, the *World Psychiatry* journal hosted a discussion on conflicts of interest in psychiatry. Most of the American psychiatrists who wrote articles had ties to the pharmaceutical industry (present or past), and while they acknowledged that such ties could lead to breaches in ethical behavior, they took care to defend the integrity of physicians who served as advisors and consultants to pharmaceutical companies. They were independent scientists, who were experts in their field, rather than "key opinion leaders." The pharmaceutical industry needed them.

"Would drug development be helped by excluding some experts from participating in industry–academic relationships and, in effect, serve as the industry watchdogs?" wrote Andrew Nierenberg, a professor of psychiatry at Harvard Medical School. "I think not. Industry needs the perspective of those physicians who best know the disorders of interest."[35]

In a similar vein, Michael Thase, a professor of psychiatry at the University of Pennsylvania School of Medicine, wrote that critics of psychiatry often failed to give proper credit "to the benefits that can result from academic-industry collaboration... Could a system that arbitrarily excluded some of the best and brightest scientists actually result in better science?"[36]

With this self-conception in mind, both Nierenberg and Thase wrote about how such "potential conflicts of interest" could be *managed*. Thase argued for "full disclosure and transparency," but what was missing from either of their papers was any sense that they personally might have been "biased" by their ties to industry. Nierenberg, for instance, disclosed that he had ties to more than a dozen pharmaceutical companies. He wrote that he was motivated to work with industry because of an altruistic desire to advance patient care. "I write this commentary as one who has evolved from rejecting any industry influence on practice (I rejected the offer of a free stethoscope from Eli Lilly during medical school) to someone who has collaborated with the pharmaceutical industry whenever I felt that it would eventually benefit patients. I believe that my relationships with industry are mutually beneficial and I value my relationships with the companies that produce the medications that help my patients."[37]

The APA, as an organization, displayed this same attitude—and same blind spot—in 2009, when it was reported that 90 percent of the panel members responsible for developing the APA clinical practice guidelines for schizophrenia, bipolar, and major depressive disorder had financial ties to pharmaceutical companies. The APA responded by telling reporters there was no reason to worry that this conflict might compromise their deliberations. "There is this assumption that a tie with a company is evidence of bias," said Darrel Regier, research director for the APA, in an interview with *USA Today*. "But these people can be objective."[38]

With that belief guiding its actions, the APA has focused on "disclosure" as a remedy for financial conflicts of interests. It required members of the *DSM-5* task force and its clinical practice guidelines to list their industry ties, and the APA also stated that those serving on the *DSM-5* task force could not receive more than $10,000 annually from drug companies. The thought was that $10,000 was too little to influence the deliberations of the committee members (which, once again, revealed an institutional blind spot to realities of implicit biases). As might be expected, the APA acted defensively when it was publicly reported that a majority of the *DSM-5* task force members had financial ties to drug companies.[39]

"We think it is important to stay focused on the fact that APA has gone to great lengths to ensure that *DSM 5* and APA's clinical practice guidelines are free from bias," said David Kupfer, chair of the *DSM-5* task force. "Throughout the development of each product, APA established, upheld, and enforced the disclosure policies and relationship limits."[40]

In this way, the APA has come to see itself as an institution that has taken the necessary steps to neutralize the influence that pharmaceutical money might have on its operations. "We adopted the strictest ethics policy of any profession for members participating in key programs such as the development of practice guidelines and the revision of DSM," APA president Jeffrey Lieberman proclaimed in 2013.[41] The public relations crisis that had arisen when Senator Charles Grassley had identified psychiatry as the most conflicted medical specialty—"our field became the poster child for physician misbehavior," Lieberman acknowledged—had passed. Indeed, Lieberman announced in 2013 that it was time for the profession to "re-engage with pharma." Pharmaceutical companies "need us and we need them," he said. The APA, through its philanthropic arm, the American Psychiatric Foundation, had already held a meeting with representatives from 14 pharmaceutical companies in order to "restore important relationships." Everyone present, Lieberman said, understood that such "interactions must be transparent, rigorously monitored, and without conflicts of interest."[42]

Without conflicts of interest—those were words that told of a profession that was ready to declare itself free from an "economy of influence," even while the monetary payments from pharmaceutical companies flowed once again.

Who Ever Said Anything about a Chemical Imbalance?

Although the APA's 1999 textbook acknowledged that the chemical imbalance theory of mental disorders had never panned out, it wasn't until 2011 that representatives of mainstream psychiatry, in various forums, began admitting that fact to the public. Those admissions, of course, raised a question for the public: If that was so, why had the public been led to believe otherwise? Why had the public been led to believe that psychiatric drugs fixed chemical imbalances in the brain? There was a sense of public betrayal in such questions, and in response, Ronald Pies, editor-in-chief of *Psychiatric Times*, which is a publication of the APA, set forth a multilayered answer.

His first blog on this subject, which we reviewed in chapter 4, bears repeating:

> I am not one who easily loses his temper, but I confess to experiencing markedly increased limbic activity whenever I hear someone proclaim, "Psychiatrists think all mental disorders are due to a chemical imbalance!" In the past 30 years, I don't believe I have ever heard a knowledgeable, well-trained psychiatrist make such a preposterous claim, except perhaps to mock it. On the other hand, the "chemical imbalance" trope has been tossed around a great deal by opponents of psychiatry, who mendaciously attribute the phrase to psychiatrists themselves. And, yes—the "chemical imbalance" image has been vigorously promoted by some pharmaceutical companies, often to the detriment of our patients' understanding. In truth, the "chemical imbalance" notion was always a kind of urban legend—never a theory seriously propounded by well-informed psychiatrists.[43]

The drug companies and enemies of psychiatry were the ones at fault. Opponents of psychiatry had falsely claimed that psychiatry had promoted the idea of chemical imbalances in order to make psychiatry look bad. Pies concluded, in that blog, that "the legend of the chemical imbalance should be consigned to the dust-bin of ill-informed and malicious caricatures."

Pies's blog stunned many readers. Did the APA really believe it had never communicated this notion to the public? In a follow-up post, Pies reiterated this stance. "Some readers felt I was trying to 'rewrite history,'" he said. "I can understand their reaction, but I stand by my statement."[44] However, he then added a new dimension to his argument: psychiatrists who told their patients that they suffered from a chemical imbalance did so thinking it would be helpful to them.

> Many patients who suffer from severe depression or anxiety or psy-chosis tend to blame themselves for the problem. They have often been told by family members that they are "weak-willed" or "just making excuses" when they get sick, and that they would be fine if they just picked themselves up by those proverbial bootstraps ... some doctors believe that they will help the patient feel less blameworthy by telling them, "You have a chemical imbalance causing your prob-lems." It's easy to think you are doing the patient a favor by provid-ing this kind of "explanation," but often, this isn't the case.

Other psychiatrists, in their public comments, sounded this same theme. They had told patients that they had a chemical imbalance because they thought it was good for them. In an interview on NPR's *Fresh Air,* Daniel Carlat explained his thinking:

> We don't know how the medications actually work in the brain. So whereas it's not uncommon—and I still do this, actually, when patients ask me about these medications, I'll often say something like, well, the way Zoloft works is it increases the levels of serotonin in your brain, in your synapses, the neurons, and presumably the reason you're depressed or anxious is that you have some sort of a deficiency. And I say that not because I really believe it, because I know that the evidence isn't really there for us to understand the mechanism. I think I say that because patients want to know something, and they want to know that we as physicians have some basic understanding of what we're doing when we're prescribing medications. And they certainly don't want to hear that a psychiatrist essentially has no idea how these medications work.[45]

In other words, psychiatrists told the chemical imbalance story to patients to give them confidence that their psychiatrists knew what they were doing. In a subsequent program on NPR's *Morning Edition,* Alan Frazer, chair of the pharmacology department at the University of Texas Health Science Center, expanded on this idea that telling patients that they had a chemical imbalance was good for them. Patients felt more comfort-able taking a drug, he said, "if there was this biological reason for being depressed, some deficiency that the drug was correcting." Telling patients they had a chemical imbalance, added University of Texas psychiatrist

Pedro Delgado, was beneficial for patients because "when you feel that you understand [your illness], a lot of the stress levels dramatically are reduced. So stress hormones and a lot of biological factors change."[46]

This was the threefold explanation that the profession offered to the public: the APA never told the chemical imbalance story to the public; enemies of psychiatry falsely claimed that the profession had done so in order to make psychiatry look ridiculous; and individual psychiatrists may have told their patients the chemical imbalance story because they thought it would be good for them. From a cognitive dissonance perspective, this explanation serves an obvious purpose. It maintains the APA's self-image as a profession knowledgeable about its research, honest in its communications to the public, and doing its best to serve the interests of patients. The fact that the public had ended up misinformed could be blamed on others, and in particular, on enemies of psychiatry intent on making the profession look bad.

However, in the world of cognitive dissonance, memories of the past are fungible, and it is easy to point out instances when the APA, including its presidents, spoke of drugs fixing chemical imbalances (as was noted in chapter 4). For instance, in a 2001 article that appeared in *Family Circle* magazine, APA president Richard Harding wrote, "We now know that mental illnesses—such as depression or schizophrenia—are not 'moral weaknesses' or 'imagined' but real diseases caused by abnormalities of brain structure and imbalances of chemicals in the brain."[47] Such examples abound. The pharmaceutical companies couldn't promote the chemical imbalance story without the tacit assent of the psychiatric profession, as our society sees academic doctors and professional organizations—and not the drug industry—as the trusted sources for information about medical maladies. Even in the summer of 2014, the APA's website, in a section titled "Let's Talk Facts" about depression, informed the public that "antidepressants may be prescribed to correct imbalances in the levels of chemicals in the brain."[48]

Equally telling, there are numerous advocacy groups that, in the summer of 2014, were still informing the public that depression was due to a chemical imbalance, and those groups have scientific advisory boards populated by prominent psychiatrists. The website of the Child & Adolescent Bipolar Foundation (recently renamed the Balanced Mind Parent Network), which lists a scientific advisory board composed of more than 20 academic psychiatrists, tells readers that "antidepressant medications work to restore proper chemical balance in the brain."[49] The Depression and Bipolar Support Alliance (DBSA) website similarly informs readers that "depression is caused by a chemical imbalance in the brain," and as it correctly informs readers, its scientific advisory board is "comprised of the leading researchers and clinicians in the field of mood disorders."[50] Michael Thase is the vice-chair of the DBSA's scientific advisory committee; David Kupfer, chair of the *DSM-5* task force, is a member of its executive committee. All told, more than 30 academic psychiatrists serve on the DBSA's scientific advisory committee. Perhaps no organization has done more to promote the chemical imbalance theory than the National Alliance on Mental Illness. Its website informs the public that "research has shown that imbalance in neurotransmitters like serotonin, dopamine and norepinephrine can be corrected with antidepressants."[51] NAMI's scientific advisory board, like that of the DBSA's, is populated by many of the most prominent psychiatrists in the country, including Nancy Andreasen, former editor-in-chief of the *American Journal of Psychiatry*, and Jeffrey Lieberman.

Research also belies the contention that telling patients they have a chemical imbalance, when they have no such known pathology, is good for them. A 2010 study by Bernice Pescosolido at Indiana University found that such explanations, if anything, increased public stigma toward the mentally ill.[52] More to the point, University of Wyoming investigators found that patients who were told they had chemical imbalances, compared to those who were not given that explanation, were more pessimistic and self-blaming regarding their prognosis. "The present findings add to a growing literature that highlight the unhelpful and potentially iatrogenic effects of attributing depressive symptoms to a chemical imbalance," they wrote.[53]

Responding to Kirsch and Fournier

The "evidence base" supporting the idea that SSRIs were wonderfully effective began to take a hit when researchers gained access to the FDA's reviews of the clinical trials for these drugs. First, there was a finding that these newer drugs had proven to be no more effective than the old tricyclic antidepressants in the trials. Then Kirsch began reporting on the "dirty little secret" contained in the FDA files, which was that SSRIs did not provide a clinically significant benefit to patients with mild, moderate, and even severe depression. It was only in the very severely ill that the drugs provided a meaningful benefit over placebo. Then there was the revelation that nearly half of the trials submitted to the FDA had failed. The FDA files told a different story than had originally appeared in the medical journals, and longer-term studies cast even further doubt on the merits of SSRIs. Rush, in his study of antidepressants in "real-world" patients, reported remarkably low response and remission rates. The results of the STAR*D trial at the end of one year—a documented stay-well rate of only 3 percent—were so bad that this finding was hidden in the published article announcing those results. Finally, as noted in an earlier chapter, in 2010 Jay Fournier and Robert DeRubeis from the University of Pennsylvania canvassed the research literature for trials of antidepressants that didn't use a placebo-washout design to suppress the placebo response, and reported that this review led to the same bottom-line conclusion as Kirsch's: "True drug effects—an advantage of antidepressant over placebo—were nonexistent to negligible among depressed patients with mild, moderate and even severe baseline symptoms, whereas they were large for patients with very severe symptoms," they concluded.[54]

Among all of these reports, it was Kirsch's findings that crystallized the issue for the public. SSRIs were the most commonly prescribed drugs in the United States. Could it be possible that, for most patients, they did not provide a clinically meaningful benefit beyond placebo? This newer information certainly contradicted what the APA had been telling the public for the past 20 years. The DART program in the 1990s stated that "recovery rates with these medications have been shown to be in the range of 70 percent to 80 percent in comparison with 20 percent to 40 percent for placebo." Which was true? Kirsch's analysis or what the APA regularly told the public? The APA and leading academic psychiatrists responded to that question in various forums, with Peter Kramer, a professor of psychiatry at Brown University, whose book *Listening to Prozac* helped kick off the Prozac era, penning a lengthy, high-profile response to Kirsch and Fournier in the *New York Times* titled "In Defense of Antidepressants."[55]

He began with a bottom-line assertion: "Antidepressants work—ordinarily well on a par with other medications doctors prescribe." Then he explained why both the Kirsch and Fournier studies could be dismissed. Kirsch, Kramer said, had "found that while the drugs outperformed the placebos for mild and moderate depression, the benefits were small." Antidepressants were still effective in that patient group, Kramer was informing the public, and then he explained why even that relatively poor result—only a small benefit—could be explained away. The problem was that "companies rushing to get medications to market have had an incentive to run quick, sloppy trials," and in their haste, they "often [enroll] subjects who don't really have depression." It was these nondepressed patients who then show up in the trial results as placebo responders, because, Kramer wrote, "no surprise—weeks down the road they are not depressed."

As even a cursory review of the facts shows, this was an explanation that Kramer had cobbled up from thin air. He wrongly reported that Kirsch had found a small drug benefit for those with mild to moderate depression (Kirsch had found that they provided no clinically significant benefit in this population), and his description of the FDA trials—that the drugs had been tested in people who really hadn't been depressed— is easily proven false. In fact, the opposite was true. In 34 of the 35 industry-funded trials that Kirsch had reviewed, the mean baseline score for patients was 23 or greater on the Hamilton Depression Rating Scale (HDRS), which is a score characteristic of "severe depression." Kirsch had segregated outcomes for the patients in those trials according to the severity of their symptoms, and he had found that it was only in those patients at the far end of this severity scale, with baseline HDRS scores above 28, that that the drugs had provided a meaningful benefit. Kramer's notion that the trials had been conducted in patients who weren't depressed at all was a figment of his imagination, although, in cognitive dissonance terms, it is very likely he believed the "truth" of his argument.

As for his critique of Fournier's and DeRubeis's study, Kramer began with what might be described as an ad hominem attack. Critics, he said, questioned "aspects of DeRubeis's math." Next, Kramer misrepresented Fournier's findings in the same way he had Kirsch's, stating that "medications looked best for very severe depression and had only slight benefits for mild depression." Once again, he had crafted a sentence that had preserved the idea that SSRIs were beneficial in mild depression, even though that was not what Fournier had found. Finally, Kramer wrote that Fournier and DeRubeis had only analyzed studies that "intentionally maximized placebo effects." In this last criticism, Kramer had turned the biased design of industry-funded trials, which use a placebo washout with the expectation it will *lower* the placebo response, into a paragon of good science. In this way, he was now arguing that trials that didn't employ such a design were "biased" in favor of placebo.

In this manner, Kramer defended antidepressants in the pages of the *New York Times*. The two studies had found that the drugs provided a small benefit for patients with mild to moderate depression, but even that finding was flawed, for a variety of reasons: the wrong patients had been recruited into trials, the trials were sloppily done, DeRubeis wasn't good at math, and the trials DeRubeis had reviewed were biased by design against the drugs. "In the end," Kramer concluded, "the much heralded overview analyses look to be editorials with numbers attached."

A short while later, *Sixty Minutes* aired its report on Kirsch's findings. This time, the APA issued an official response, which took Kramer's defense a step further. To claim that there was no difference between antidepressants and placebos, said APA president John Oldham, was "not just wrong, but irresponsible and dangerous reporting." Added Jeffrey Lieberman, who was then president-elect of the APA: "Kirsch has badly misinterpreted the data and his conclusion is at odds with common clinical experience. He has communicated a message that could potentially cause suffering and harm to patients with mood disorders."[56]

The APA's message was clear: antidepressants worked and *Sixty Minutes* should be ashamed of itself for suggesting otherwise. The venerable television program had failed to report what everyone knew to be true.*

* In the APA's release, there was also this cognitive dissonance gem: "The APA's Treatment Guidelines on depression recommend psychotherapy first for mild to moderate depression, and only after this approach falls short should the physician decide whether or not antidepressants are needed in conjunction with psychotherapy." This is not true (see chapter seven), but it seems the APA here was intent on presenting an image of an organization that was not in thrall to drugs. And it is a statement that makes it seem as though the APA's clinical practice guidelines are consistent with Kirsch's findings, which it denounced in the same press release as worse than wrong.

Kill All the Critics

In settings that provoke cognitive dissonance, one common response by
the threatened person (or organization) is to question the integrity of the
critic. Tavris and Aronson describe this as a "kill the messenger" response.
The critic is biased, or has poor math skills, or is horribly misinformed, or
is simply against psychiatry, eager to make the profession look bad. There
are numerous examples of the APA behaving in this way when findings
threatened the self-image of the profession, and while it isn't surprising
that the APA has directed that response toward lesser-known figures in
the scientific community, such as Kirsch and Fournier, it is notable that
the organization has responded in this way even to those of high stature
within medicine.

In 2011, Marcia Angell, former editor-in-chief of the *New England
Journal of Medicine*, wrote a two-part essay for the *New York Review of Books*
that provided a review of three books that were critical of psychiatry.*
One of the two parts was titled "The Illusions of Psychiatry," and in the
essay she wrote of the spurious nature of the chemical imbalance story,
the many failings of the DSM, Kirsch's work, and the lack of evidence
showing that psychiatric drugs provide a long-term benefit. The three
books, she concluded, "are powerful indictments of the way psychiatry is
now practiced."[57]

Now the *New England Journal of Medicine* is one of the most prestigious
medical journals in the world. As the former editor-in-chief of that journal, Angell had impeccable credentials within the medical world—her
expertise in evaluating the quality of clinical research could not possibly
be questioned. Yet, that is what the APA and several prominent psychiatrists did in response. In *Psychiatric News,* the APA called her essay "simplistic at best and deeply misinformed at worst," with APA president John
Oldham stating "there's a lot of very bad distortion in here for someone
with her stature to be promoting."[58] Oldham also fired off a "letter to the
editor" of the *New York Review of Books,* writing that the APA regretted
that Angell had not taken "a more balanced approach," and asserting—
much as Kramer had done in his response to Kirsch—that today, "thanks
to medical and therapeutic advances, there is real help for those who suffer
the devastating effects of mental illness."[59] Next, Darrel Regier authored
an "official APA response" to her essay, chiding Angell for "showing no
appreciation of the scientific advances that have been made in the understanding of mental disorders in recent decades"; for her "pontification" on
psychiatry's failures; and for her "superficial treatment of psychiatry" in
her "attempted review of three disparate books."[60]

---

* The three books were *The Emperor's New Drugs: Exploding the Antidepressant Myth* by Irving
Kirsch; *Anatomy of an Epidemic: Magic Bullets, Psychiatric Drugs, and the Astonishing Rise of Mental Illness
in America* by Robert Whitaker; and *Unhinged: The Trouble with Psychiatry—A Doctor's Revelations
About a Profession in Crisis* by Daniel Carlat.

Several prominent psychiatrists also wrote letters to the editor of the *New York Review of Books*, including Richard Friedman from Weill Cornell Medical College and Andrew Nierenberg from Massachusetts General Hospital. They criticized Angell for using "an outdated and disproven chemical imbalance theory of depression" as a "straw man" to criticize psychiatry. They described Angell as "unaware of recent advances in neuroscience research," and claimed that she had "distorted the potential adverse effects of psychotropic drugs with anecdotes and flawed data and downplayed the devastating consequences of untreated psychiatric illness." They concluded "it would be sad—and harmful—if any patients were discouraged from seeking safe and effective psychopharmacological treatment on the basis of Angell's uncritical and biased review."[61]

Parse those responses and you find a remarkable personal attack: Angell was biased, superficial, uncritical, unaware, and deeply misinformed, and she had engaged in "distortions" that could bring great harm to patients. The APA and the academic psychiatrists weren't simply disputing Angell's review with these comments; there was a snide, dismissive tone to their words. And such was the response to someone who, as Angell noted in her response to their letters, had spent her life "evaluating the quality of clinical research."

The APA also employed this "kill the messenger" card when two of its own, Robert Spitzer and Allen Frances, chair of the *DSM-IV* task force, criticized the *DSM-5* task force for its lack of transparency and other sins. Spitzer was upset that members of the *DSM-5* task force were required to sign a confidentiality agreement that prevented them from disclosing anything about their deliberations. He believed this would compromise the scientific integrity of the process. Frances voiced similar concerns, writing that the creation of *DSM-5* had turned into a "closed and secretive" process.[62] In July of 2009, Spitzer and Frances wrote to the APA's board of Trustees, and this time they upped the ante: they were worried that the *DSM-5* task force was going to expand diagnostic boundaries in a way that would "add tens of millions of newly diagnosed 'patients'—the majority of whom would likely be false positives subjected to the needless side effects and expense of treatment."[63] The APA's response, authored by Alan Schatzberg, James Scully, David Kupfer, and Darrel Regier, concluded in this way: "Both Dr. Frances and Dr. Spitzer have more than a personal 'pride of authorship' interest in preserving the *DSM-IV* and its related case book and study products. Both continue to receive royalties on *DSM-IV* associated products. The fact that Dr. Frances was informed at the APA annual meeting last month that subsequent editions of his *DSM-IV* associated products would cease when the new edition is finalized, should be considered when evaluating his critique and its timing."[64]

Spitzer and Frances, the APA was now arguing, were motivated by their own personal conflicts of interest. These two titans of psychiatry, now that they were raising such criticism, deserved to be lumped with Angell and all the others who, for one reason or another, failed to see modern psychiatry through properly appreciative eyes.

Mistakes Were Made, But Never by Us

This exploration of cognitive dissonance theory—and the APA's response to dissonant settings—provides a sense of what is possible, in terms of setting forth prescriptions for reform. And what is evident in this chapter is that there is little possibility that psychiatry, as an institution, can see a need for such reform, or even see, in any substantial way, that it has been corrupted by the twin economies of influence discussed in this book. Ever since the publication of *DSM-III*, the APA has cultivated a self-image, which it has presented both to the public and to itself, of a medical specialty that has made great progress in discovering the biology of mental disorders, and one that has, in its medical armamentarium, an array of drugs that are quite effective in treating mental disorders. The response by the APA and leaders of the field to research findings or criticisms that challenge that self-image has regularly been to protect it at all costs. Researchers of cognitive dissonance will recognize the usual mechanisms employed: assertion of one's personal integrity, manipulation of "facts" that resolves the dissonant state, declaration that the critics are just plain wrong, and the denigration of critics.

Tavris and Aronson wrote a book titled *Mistakes Were Made (But Not by Me)*. That is an apt phrase to describe psychiatry's look back at its own past.

Psychiatry's insecurity as a medical discipline undoubtedly contributes to this mind-set within the field. The APA has worked diligently over the past 30 years to improve its image with the public, and given the story it has told to the public over this period, which has regularly emphasized remarkable advances, it can hardly now be expected, in its own "mind," to admit—as Marcia Angell's essay noted—to its own "illusions." Indeed, in one of his last acts as president of the APA, Jeffrey Lieberman wrote an article for *Scientific American* that cast psychiatry's struggles with its most ardent critics as a battle, really, between good and evil. Psychiatry, he said, had to confront "antipsychiatry" groups that "have relentlessly sought to undermine the credibility of psychiatric medicine and question the validity of mental illness."[65]

> These are real people who don't want to improve mental healthcare, unlike the dozens of psychiatrists, psychologists, social workers and patient advocates who have labored for years to revise the DSM, rigorously and responsibly. Instead, they are against the diagnosis and treatment of mental illnesses—which improves and in some cases saves, millions of lives every year—and "against" the very idea of psychiatry, and its practices of psychotherapy and psychopharmacology. They are, to my mind, misguided and misleading ideologues and self-promoters who are spreading scientific anarchy...This relatively small "antipsychiatry" movement fuels the much larger segment of the world that is prejudiced against people with disorders of

the brain and mind and the professions that treat them. Like most prejudice, this one is largely based on ignorance or fear—no different than racism, or society's initial reactions to illnesses from leprosy to AIDS. And many people made uncomfortable by mental illness and psychiatry, don't recognize their feelings as prejudice. But that is what they are.

This is a mind-set that undermines the self-examination, or introspection, needed to develop a framework for substantial reform.

CHAPTER ELEVEN

Prescriptions for Reform

Those of us who indict past failures have a duty to develop new solutions.

—Devra Davis, 2009[1]

As we noted in the introduction, this book grew out of a year we spent as fellows in a lab on institutional corruption at the Edmond J. Safra Center for Ethics. The focus there was always solution oriented, as opposed to blame oriented. Any investigation of institutional corruption should illuminate the systemic influences that have led the institution astray, and set forth prescriptions for reform. We have now arrived at this ultimate step: How can psychiatry be reformed? Or—and this is a more provocative question—*can* it be reformed? Perhaps psychiatry's place in our society, as the ultimate authority in this domain of mental health, needs to be rethought.

Even as we are setting out to write this chapter, we are not certain of our own answers to these questions. However, the framework of institutional corruption is meant to provide those studying the problem with a *process* for thinking about solutions. In addition, any reform effort must necessarily involve the public, and thus, if we can lay bare this process, readers can join in and participate in thinking about what might be done. There are no simple right answers. But we are certain that there is a need for the public to develop a remedy to the institutional corruption that we have documented in this book, which is causing such harm to our society.

Avoid the Fundamental Attribution Error

When we learn about corrupt behavior that is harming our society, we naturally tend to think of how it involves "bad actors" doing "bad things." The individual's corrupt behavior is seen as arising from an internal character flaw, rather than in response to external, or situational, influences.

In academic circles, that is known as a "dispositional" explanation for the corruption. However, in order to develop prescriptions for reform, it is essential that we shift away from such "dispositional" explanations (e.g., the problem is that corrupt psychiatrists hid the truth) to *contextual* and *systemic* explanations. Whereas dispositional explanations nurture a sense of public outrage and a desire to blame individuals, avoiding the "fundamental attribution error" means appreciating the "potent, though often unnoticed influences of the situation."[2] This latter approach naturally fosters thinking about what can be done to change the systemic influences that are at the root of the corruption.

Physician Benjamin Falit has written about this need to avoid the fundamental attribution error in his exploration of ways to motivate industry to conduct post-approval drug trials that are "optimally informative to the medical community."[3] He observes that pharmaceutical companies should be seen as "corporate actors enmeshed within a complex regulatory framework," and thus, within that framework, they naturally seek to conduct trials that help market their drugs (and CEOs of publicly traded companies also have a fiduciary duty to their shareholders, which means seeking to maximize profits). If the public wants to enjoy the benefit of postapproval trials that are "optimally informative to the medical community," then it will need to amend the regulatory framework governing such research. The public cannot expect drug companies to simply choose to alter their behavior to serve the public good.

If we look at organized psychiatry in this way, we can immediately see the many systemic influences that have affected its behavior. In chapter 2, we detailed the situational influences on psychiatry during the 1970s, when it chose to remake its diagnostic manual and embrace a "medical model." At that time, the profession felt that it was "under siege," and that its standing in society—as a medical specialty—was being challenged. The profession also found itself in competition with psychologists, social workers and other therapists for care of those with milder emotional difficulties (the "walking wounded"). The APA had scientific reason to rethink its diagnostic manual, but those cultural influences were also present. Once *DSM-III* was published, the pharmaceutical industry's financial influence over the APA and academic psychiatry became pronounced. In addition, psychiatry had a need, as a guild, to promote its new medical model to the public, and that guild interest was further strengthened by psychiatry's own fragile psyche. There are a multitude of financial, cultural, and psychological factors that have shaped the behavior of the APA and academic psychiatrists over the past 35 years.

At the same time, cognitive dissonance theory tells us that we should not expect that individuals within organized psychiatry would be aware that their behavior may be ethically compromised. Behaviors that are seen by outsiders as unethical may have become normative within the institution; in addition, psychiatrists' self-image almost necessarily precludes such self-awareness. Understandably, they see themselves as physicians who put their patients' interests first. Those who participate in research see themselves as scientists (or at least scientific in their thoughts and methods). Max Bazerman, a business ethicist at Harvard University, uses the terms "ethical blind spots" and "bounded ethicality" to describe the limitations of the conscious mind in addressing information that may challenge one's sense of self.[4] Psychiatrists, like nearly all of us, are motivated to eradicate cognitive dissonance through self-serving bias and rationalization.

In a sense, "avoiding the fundamental attribution error" requires that those who would prescribe remedies for institutional corruption take an imaginative leap. In order to truly appreciate the corrupting nature of "economies of influence," it is helpful to imagine one's own self in that environment. In this particular instance of institutional corruption, it became acceptable practice for academic psychiatrists to serve as speakers, advisors, and consultants to pharmaceutical companies. Psychiatrists were also incentivized to engage in behaviors, such as putting their names on ghostwritten papers, that ran counter to psychiatry's public health mission. It became normal for academic psychiatrists and the APA to defend guild interests at every turn. By imagining one's own self within that environment, we can better focus on the systemic influences that led the field astray, and avoid the impulse to think of "dispositional" explanations for the corruption we have detailed in this book.

Making the Problem Known

Institutional corruption causes harm to a *society*. In order for reform to take place, it is essential that society become aware of the corruption, the reasons for it, and the harm it is causing. The society can then seek to eliminate, or at least neutralize, the corrupting influence. The institution is understood to be operating in an environment that led to corrupt behavior, and thus society must propose reforms that will change that environment.

As part of this process, of making a problem known through this lens, there is the hope that it will enable more individuals within the institution to see the need for reform. The thought is that if the explication of the corruption within this framework is done well, then it may enable "insiders" to recognize their own cognitive dissonance. If that occurs, there is a real foundation for change, with the institution seeking to reform itself even as society seeks to enact corrective measures. However, as we saw in the last chapter, the APA is already taking steps to "re-engage with pharma," and is doing so with little apparent appreciation that psychiatry was compromised in a significant way by its close ties to industry. Nor is there any evidence that the APA and academic psychiatry are aware that guild influences have proven to be so corrupting. As such, it seems unlikely that substantial reform will arise from within psychiatry. The burden will be on society to develop the needed reform measures.

Transparency Is Not a Solution

The medical profession has turned to disclosure of financial conflicts of interest as a remedy for neutralizing their corrupting effects. The thought is that if the financial tie is made transparent, then this will diminish or eliminate the bias that may arise from the conflict. Those who favor disclosure also often argue that it would be counterproductive to eliminate such financial ties between academic researchers and industry. This collaboration is needed to improve patient care. As Thomas Stossel, from Brigham and Women's Hospital in Boston, wrote: "How could unrestricted grants, ideal for research that follows up serendipitous findings, possibly be problematic? The money leads to better research that can benefit patients."[5]

The APA has adopted this approach. Transparency and disclosure of commercial ties are seen as a remedy for financial conflicts of interest. As noted in previous chapters, the APA required *DSM-5* panel members to disclose any commercial ties they had had during the three years prior to their serving on the DSM project, and stated that they were prohibited from having current ties that paid them more than $10,000 annually. The APA adopted other disclosure measures, and, as was noted in the last chapter, it now prides itself as a medical discipline that has taken a leading role in disclosing financial ties.

The problem with disclosure as a proposed solution is that it doesn't eliminate the bias that may arise from the financial relationship. The conflict remains, and research shows that disclosure may, in fact, worsen the implicit bias. Moreover, transparency and disclosure will not restore public trust in the psychiatric profession. For instance, what are primary care physicians to think when they turn to the APA's clinical practice guide for depression and see that *all* of the authors of the guideline disclosed financial ties to pharmaceutical companies that manufacture antidepressants? Do the physicians think that since the ties are disclosed, there is nothing to worry about? That such ties are evidence of the authors' expertise and thus the recommendations can be trusted? Or do they think that such ties likely bias the authors' recommendations, and thus the guideline might not be valid and trustworthy? And what was the public to think when they learned that a majority of the members of the *DSM-5* panels had ties to pharmaceutical companies? That this wouldn't affect their setting of diagnostic boundaries? The pharmaceutical companies benefit from the expansion of diagnostic boundaries, and the public drew the obvious conclusion, which was that commercial interests might have influenced the *DSM-5* panel members in their deliberations. All disclosure of financial ties does is make the public more readily aware of the conflicts of interest that bedevil the profession.

The disclosure of industry ties also does not do anything to address the conflict of interest that arises from guild interests. That economy of influence remains, with disclosure simply serving to direct everyone's attention toward the industry problem, leaving this one—which may be the more powerful influence of the two—mostly out of sight.

Neutralizing Industry Influence

Since disclosure is not a solution to preventing bias, one obvious potential remedy is to simply eliminate financial conflicts of interest among those who are on DSM panels, CPG committees, and involved in writing APA-published textbooks. In particular, those who staff such panels and committees should not be serving as consultants, advisors, or speakers for pharmaceutical companies (serving as an investigator in a clinical trial of a new drug is not necessarily compromising, however). The APA could then present its work in those key domains as free from any direct industry influence.

The thought of eliminating such ties is not a new idea. However, the APA, like many medical specialty organizations that produce their own diagnostic and treatment guidelines, has maintained that it is not possible to find enough psychiatrists with the requisite expertise for such tasks who don't have commercial ties. While that may have been true 20 years ago, it is less so today, and the pool of conflict-free academic psychiatrists should continue to grow.[6]

By the mid-1990s, a very high percentage of academic psychiatrists had financial ties to industry, so much so that when the *New England Journal of Medicine* tried to find an expert to write about antidepressants, it had difficulty finding anyone who didn't have such ties. It is easy to understand why such ties were omnipresent: the academic psychiatrists benefitted in multiple ways from them.

The pharmaceutical companies paid them well, and these ties also helped them prosper in their academic careers. Their work for pharmaceutical companies often brought in research funds to their medical schools, and even more important, it helped them become known as national and international experts. Those with ties became "thought leaders" within the profession. Thus, serving as a consultant or advisor to drug companies brought financial reward, career advancement, and societal prestige. There was little personal downside. However, that risk–benefit equation began to change when the public began to regularly hear of how pharmaceutical money was corrupting medicine. Society started to question whether those researchers could be neutral and objective, and this changed the cost–benefit analysis for an individual deciding whether to become an advisor or consultant to a pharmaceutical company. The monetary rewards from industry came to be understood as having a societal—and professional—cost. Indeed, in this sense, it can be argued that disclosure does work: it provides an incentive for researchers to free themselves from such ties.

As a result, there are now a growing number of academics, in psychiatry and other medical specialties, who do not serve as consultants, advisors or speakers to pharmaceutical companies (and some who don't serve as investigators in industry-funded trials). These academics also make this known; this becomes a "disclosure" that reflects well on them. Academic medical centers could further encourage this movement by reserving endowed chairs and other prestigious academic positions for faculty without such ties. "If greater academic prestige accrues to distant rather than close relationships with industry, a new social norm may emerge that promotes patient care and scientific integrity," write Sunita Sah and physician Adrian Faugh Berman.[7]

When the APA, under Jeffrey Lieberman's leadership, decided in 2013 that the profession should take steps to renew its ties to industry, it was missing an opportunity. A medical specialty that develops its own diagnostic manual and CPGs has a more valuable product if such work is developed by academic experts with no ties to industry. The absence of such conflicts, when disclosed, would present the APA and its work in a better light. In accounting terms, there is a "good-will" value associated with a product of that sort, but the APA has failed to recognize that. For instance, when the public learned that more than 50 percent of the *DSM-5* committee members had financial ties to industry, the APA naturally found itself on the defensive. The DSM panel members with such ties could still be objective, the APA said. For the public, that was a response that triggered a roll of the eyes, and therein lies the cost to the APA—a diminishment of societal trust in the profession.

We wrote in the beginning of this chapter that blaming individuals would not be a productive path toward a solution. But in this instance, societal criticism of such ties serves as a powerful force for change. The young academic psychiatrist eager to assume a leading role in the profession now has to weigh that cost against the financial reward, and may conclude that developing financial ties to industry will hamper pursuit of that career goal. As for the APA, its leaders often complain that psychiatry stands alone among the medical professions in having to confront critics who question its legitimacy. With societal criticism of industry–academic ties having become pronounced, the APA should understand that its standing in society would be enhanced if it required that its DSM and CPG committees be staffed by academic psychiatrists without ties to industry.

The APA could also enhance its stature if it eliminated such conflicts in other areas of its operations. It could require that its officers be free from such ties. It could institute a policy that barred CME offerings funded by industry. It could decide that the board members of its two philanthropic arms—the American Psychiatric Institute for Research (APIRE) and the American Psychiatric Foundation—could not have such ties to industry. In 2010, for instance, it was reported that 56 percent of APIRE's board were drug industry executives, with one member the CEO for marketing for Eli Lilly.[8] The presence of industry executives on that board taints that enterprise.

The important thing to understand is that the APA cannot be expected to make such changes because it is the "right" or "ethical" thing to do. However, if societal attitudes become so pronounced on this issue, such that the public becomes insistent that it wants medical specialties to be led by academic researchers who do not have financial ties to industry (other than perhaps a research tie), then that will serve as an "economy of influence" that will push the APA in a new direction. The elimination of such ties will simply make good sense from a *guild* perspective.

A Financial Conflict of a Different Sort

The APA and pharmaceutical companies have rightfully argued that it would be a loss if academic psychiatrists didn't serve as investigators in clinical trials of new drugs. The question then becomes how can that scientific task be regulated so that the academic investigators, rather than lending their names and prestige to an industry-controlled process, are in a position to *independently* test the new drugs. This would remove the industry "economy of influence" on this process (although the guild interest might still be present, as will be explored below).

As society seeks regulatory reform that could achieve this goal, it is useful to note that industry's assertion of control of its studies is of fairly recent origin. Up until the mid-1980s, pharmaceutical companies seeking to test their new drugs had to come to academic investigators with their "hat in hand." They had nowhere else to go, and the academic researchers, who pursued National Institute of Health grants to fund their research, were often reluctant to conduct the trials. That was a market dynamic that gave power to the academic researchers. When they agreed to conduct the research, they would insist on intellectual control: they would design the trials, collect and analyze the data, and author their own papers.

However, that dynamic soon changed. The rise of health maintenance organizations curbed the earnings of community physicians, and many physicians began looking for other opportunities to supplement their incomes. That's when the drug companies came knocking on their doors. These physicians did not have the knowledge or skills to lead clinical drug trials, but they could be trained to follow a drug protocol and enroll patients. The drug companies no longer needed the academic physicians as much, since they could get most of the patients needed for a trial through community doctors. Thanks to this new environment, they could now say to the academic physicians, if you want our research dollars, you have to participate on our terms. This new message came at the very time that academic physicians were finding it more difficult to obtain NIH grants, and also at a time that their medical schools, for various reasons, were trying to increase their faculty's participation in industry-funded research.

The negotiating power had swung to the pharmaceutical companies, and they exploited that new leverage by asserting control of the trials: they designed the studies, analyzed the data, and, with great frequency, hired medical writing firms to prepare articles on study results. The academic investigators would then be given a chance to review the draft manuscripts and ask for changes before signing off as authors on the published papers.

The challenge for society, then, is to create a regulatory environment for the testing of new drugs that restores a scientific integrity to this process. For example, fellows at the Safra Center have shown that "blinding" of the funding source—for example, when research funds are pooled and the researchers don't know who is funding their work—may be an effective and practical solution.[9] Many scholars are now developing proposals of this sort, and detailing them is beyond the scope of this book. We are simply noting that whereas it is important that academic psychiatrists serving on diagnostic and clinical practice guideline panels not act as consultants, advisors or speakers for drug companies, their participation in industry-funded trials is a problem of a different sort. The reform that is needed wouldn't keep them from participating in such trials, but rather would ensure that they—perhaps in concert with researchers from other disciplines—had intellectual control of the studies.

Neutralizing the Guild Interest

For some time, society's attention has been focused on the corrupting influence of pharmaceutical money on psychiatry and all of medicine. Much less attention has been given to the corrupting influence of psychiatry's guild interests, which, as we have seen in this book, is the deeper, more intransigent influence. How can it possibly be negated? There is only one real solution that we can see: society needs to require psychiatry to share authority over "psychiatric problems" with other parties, such that its voice becomes one among many, rather than the ruling one, which it has been for the past 35 years.

Psychiatry's authority in this realm arose in two steps. The first came in the mid-1800s, when physicians stepped in as superintendents of the early moral therapy asylums. This led to the birth of psychiatry as a medical specialty, and psychiatry has retained its societal authority over the "mad" ever since. Its authority over the "walking wounded"—the second step in psychiatry's rise to its prominent place in society today—was established when the APA published *DSM-III*. As was seen in chapter 2, during the 1970s psychiatry was in competition with numerous others—psychologists, social workers, counselors, and various other therapists—for this group of patients, who sought help for the multitude of emotional struggles that can afflict human beings. When psychiatry adopted its medical model, it was asserting its authority over this group of patients, and it was doing so by trading on its social stature as a medical discipline. If depression, anxiety, and other common emotional difficulties were illnesses, or "diseases of the brain," then it made sense that physicians would have authority over these illnesses.

The challenge for psychiatry following the publication of *DSM-III* was to secure its authority over this realm. It took immediate steps to do so, setting up a public relations effort to sell this new model to the public, an effort it has continually maintained. Other medical specialties don't need to conduct PR campaigns because their authority, in their particular disease realm, is well established. But psychiatry has needed to do so precisely because its authority over such a wide swath of American society lacks a solid scientific foundation.

Given this vulnerability, it is easy to understand why the APA and academic psychiatry have always felt a need to protect the field's medical model. Psychiatry's authority is dependent on it. Thus, when scientific inquiries failed to validate that model, or produced results at odds with the story that the APA's PR machinery had been telling, guild influences naturally came into play. For instance, when the field trials of *DSM-5* produced poor kappa scores, could the authors of that manual (and the APA) promote a new manual that lacked diagnostic reliability? If that was so, and if, as NIMH director Thomas Insel had said by this time, the DSM categories also lacked validity, then how could *DSM-5* be presented to the public as a useful manual? Lowering the standards for what could be considered a good or acceptable kappa score became a way for the *DSM-5* authors to maintain a claim that its new manual had some scientific merit. The poor results from the NIMH trials of stimulants and antidepressants presented the field with a similar challenge. Those studies did not produce results that told of effective long-term treatments for known diseases. Could the profession really afford to highlight such findings? The reason that guild influences were so powerful at these moments is precisely because the stakes were so high for psychiatry: the results did not just tell of a flawed treatment, or of a single mistaken diagnosis, but rather raised questions about the merits of the "medical model" that provided psychiatry with its authority in this realm.

This is why it is necessary for society to fundamentally rethink psychiatry's authority over psychiatric care. We need a new model for thinking about psychiatric issues, and we need authority for such care to be distributed among a broader group of professionals. That is the only way to negate the guild "economy of influence" that has proven so corrosive to American society over the past 35 years.

A Paradigm Shift

The leaders of the APA often speak about adopting a bio-psycho-social model of care. That is a fine idea, particularly when the word is carefully parsed: bio + psychological + social. Psychiatry might lay claim to being the experts in the bio part of that equation, but other professionals could lay claim to being experts in the other two parts. Psychiatry has no special expertise in psychological or social matters. Truly adopting a bio-psycho-social model would mean that care in this arena should be led by a broad collection of people: psychiatrists, psychologists, social workers, philosophers of the mind, sociologists, and so forth. This broader mix of "authorities" could then deliberate over diagnostic categories, or whether such "labeling" is even helpful, and deliberate over the multitude of therapeutic approaches that might be of use.

This could dramatically alter our societal thinking about how to promote mental health. For instance, if a significant percentage of youth are having difficulty focusing in the classroom, perhaps it is the classroom environment that needs to be changed. Our society would also need to look at how poverty, injustice, and stressful work environments can cause depression, anxiety, and other emotional turmoil. Remedies for such psychological difficulties might include the selective use of psychiatric drugs, as well as nutrition, exercise, social activities, and other lifestyle changes. The NIMH could also direct a portion of its research dollars toward the assessment and development of such holistic care.

There are many critics of psychiatry, both from within the profession and outside it, who are envisioning a different future for psychiatric care. For example, Irish psychiatrist Pat Bracken and English psychiatrist Philip Thomas have written about a "post-psychiatry" future. For so long, they note, societal discussion on this topic has been characterized by a polarized conflict between psychiatry and "antipsychiatry" groups. They argue for a psychiatry informed by a multitude of voices, with the voice of the "service user" given particular import. "Post psychiatry does not propose new theories about madness, but opens up new spaces in which other perspectives can assume a validity previously denied them," they write.[10] In this post-psychiatry future, new possibilities would appear.

The time is clearly ripe for a paradigm shift. There are many in American society who are seeking alternatives to psychiatry's medication-centered care, and they are turning to nutrition, exercise, meditation, and other pursuits to get well and stay well. Meanwhile, from a scientific standpoint, psychiatry is clearly facing a legitimacy crisis. The chair of *DSM-IV,* Allen Frances, has written about the disgrace of the *DSM-5* field trials, with the APA maintaining, in spite of the data, that the kappa scores showed acceptable reliability. The chemical imbalance theory is collapsing now in the public domain. Insel and others are writing about how the second generation of psychiatric drugs is no better than the first, which belies any claim that psychiatry is progressing in its somatic treatments of psychiatric diseases. The burden of mental illness in our society, among children and adults alike, continues to increase. The disease model paradigm embraced by psychiatry in 1980 has clearly failed, which presents society with a challenge: what should we do instead?

A study of institutional corruption is supposed to illuminate the social factors that led to the corruption, detail the resulting social injury, and provide prescriptions for reform. Perhaps readers of this book will be able to imagine reforms that neutralize the guild "economy of influence" that has led psychiatry astray while still preserving its authority in this realm. However, we—the two authors of this book—think that a more fundamental reform is needed. The remedy that is needed is to eliminate psychiatry's hegemony over this domain of American life. That is a societal function too important to leave in the hands of an institution buffeted so strongly by guild interests. Instead, societal authority needs to be invested in a broader group of professionals (and thinkers), who collectively could be said to provide expertise in biology, psychology, and sociology. In the absence of a dominant guild influence, this "biopsychosocial group" could be expected to hold true to its public-service mission. And if that were so, our society could hope to benefit from psychiatric care quite different from the diagnostics and treatments we have known in the recent past.

NOTES

1 A Case Study of Institutional Corruption

1. Jonathan H. Marks. "Instrumental Ethics, Institutional Corruption, and the Biosciences," Conference on the Future of the Humanities, Amherst College, Amherst, MA. March 3, 2012.
2. D. Wikler. "A Crisis in Medical Professionalism," in *Ethics and the Business of Biomedicine*, edited by D. Arnold (New York: Cambridge University Press, 2009): 253.
3. G. Fava. "Financial conflicts of interest in psychiatry." *World Psychiatry* 6 (2007): 19–24.
4. M. Angel. "Is academic medicine for sale?" *New Engl J Med* 342 (2000): 1516–18.
5. K. Quanstrum. "Lessons from the mammography wars." *New Engl J Med* 363 (2010): 1076–9.
6. L. Lessig. "What an originalist would understand 'corruption' to mean." *California Law Review* 102 (2014): 1–24.
7. Medco. "America's State of Mind." November 2011.

2 Psychiatry Adopts a Disease Model

1. R. Mayes. "DSM-III and the revolution in the classification of mental illness." *J Hist Behav Sci* 41 (2005): 249–67.
2. IMS Health. "Top Therapeutic Classes by U.S. Spending," 2007–2011. Spending on antipsychotics, antidepressants, and ADHD medications in the U.S. totaled $37.1 billion; spending on "anti-epileptics" for treatment of mood disorders is estimated to have comprised at least $3 billion of the $5.9 billion spent on this class of drugs in 2011, bringing the total spending in 2011 to above $40 billion.
3. G. Grob. "Origins of DSM-I: a study in appearance and reality." *Am J Psychiatry* 148 (1991): 421–31. In this article, Grob provides a detailed review of the history of psychiatry nosology that led up to the APA's creation of *DSM-I*; we have relied on his account for this brief description of psychiatry's nosology prior to *DSM-I*.
4. Nancy Tomes, *The Art of Asylum Keeping* (University of Pennsylvania Press, 1994): 124.
5. Grob, ibid.
6. Grob, ibid.
7. Grob, ibid.
8. Grob, ibid.

9. Grob, ibid.

10. Grob, ibid.

11. H. Decker. *The Making of DSM-III* (New York: Oxford University Press, 2013): 4.

12. Grob, ibid.

13. M. Wilson. "DSM-III and the transformation of American psychiatry: a history." *Am J Psychiatry* 150 (1993): 399–410.

14. Wilson, ibid.

15. Grob, ibid.

16. A. Beck. "Reliability of psychiatric diagnoses." *Am J Psychiatry* 119 (1962): 351–7.

17. R. Spitzer. "A re-analysis of the reliability of psychiatric diagnosis." *Br J Psychiatry* 125 (1974): 341–7.

18. D. Rosenhan. "On being sane in insane places." *Science* 179 (1973): 250–8.

19. Wilson, ibid.

20. K. Kendler. "The development of the Feigner criteria: a historical perspective." *Am J Psychiatry* 167 (2010): 134–42.

21. Kendler, ibid.

22. "Wonder Drug of 1954?" *Time*, June 14, 1954.

23. "Don't-Give-a-Damn Pills." *Time*, February 27, 1956.

24. M. Strand. "Where do classifications come from? The DSM-III, the transformation of American psychiatry, and the problem of origins in the sociology of knowledge." *Theor Soc* 40 (2011): 273–313.

25. J. Maciver. "Patterns of psychiatric practice." *Am J Psychiatry* 115 (1959): 692–7.

26. T. Szasz. *The Myth of Mental Illness,* rev. edition (New York: Harper and Row, 1974; first edition 1960): xiii.

27. See blurbs for second edition of *The Myth of Mental Illness*, published by Harper and Row in 1974.

28. R. Buchanan. "Legislative warriors: American psychiatrists, psychologists, and competing claims over psychotherapy in the 1950s." *J Hist Behav Sci* 39 (2003): 225–49.

29. Buchanan, ibid.

30. R. Mayes. "DSM-III and the revolution in the classification of mental illness." *J Hist Behav Sci* 41 (2005): 249–67.

31. Strand, ibid.

32. Strand, ibid.

33. Wilson, ibid.

34. Wilson, ibid.

35. T. Hackett. "The psychiatrist: in the mainstream, or on the banks of medicine." *Am J Psychiatry* 134 (1977): 432–4.

36. Wilson, ibid. Hackett, ibid.

37. Decker, ibid, 141.

38. Decker, ibid, 142.

39. Wilson, ibid.

40. M. Sabshin. "On remedicalization and holism in psychiatry." *Psychosomatics* 18 (1977): 7–8.

41. S Kety. "From rationalization to reason." *Am J Psychiatry* 131 (1974): 957–63.

42. A. Ludwig. "The medical basis of psychiatry." *Am J Psychiatry* 134 (1977): 1087–92.

43. S. Guze. "Nature of psychiatric illness: why psychiatry is a branch of medicine." *Comp Psychiatry* 19 (1978): 295–307.

44. F. Redlich. "Trends in American mental health." *Am J Psychiatry* 135 (1978): 22–8.

45. R. Spitzer. "Research diagnostic criteria: rationale and reliability." *Arch Gen Psychiatry* 35 (1978): 773–82.

46. Wilson, ibid.
47. Decker, ibid, 145.
48. Decker, ibid, 58.
49. A. Spiegel. "The Dictionary of Disorder." *The New Yorker*, January 3, 2005.
50. Strand, ibid.
51. Spiegel, ibid.
52. C. Dean. "Diagnosis: the Achilles' heel of biological psychiatry." *Minnesota Medicine* 74 (1991): 15–17.
53. Mayes, ibid.
54. Decker, ibid, 252.
55. Decker, ibid, 257.
56. D. Adler. "The medical model and psychiatry's tasks." *Hosp Community Psychiatry* 32 (1981): 387–92.
57. Mayes, ibid.

3 Economies of Influence

1. L. Lessig. "We the People, and the Republic We Must Reclaim." TED talk, February 2013.
2. M. Sabshin. "Report of the medical director." *Am J Psychiatry* 137 (1980): 1308–12.
3. L. Havens. "Twentieth-century psychiatry." *Am J Psychiatry* 138 (1981): 1279–87.
4. M. Sabshin. "Report of the medical director." *Am J Psychiatry* 138 (1981): 1418–21.
5. Sabshin, ibid.
6. W. Sorum. "Report of the speaker." *Am J Psychiatry* 140 (1983): 1403–4.
7. Sabshin, 1981 report of the medical director, ibid.
8. Sabshin, ibid.
9. M. Sabshin. "Report of the medical director." *Am J Psychiatry* 139 (1982): 1392–6.
10. M. Sabshin. "Report of the medical director." *Am J Psychiatry* 140 (1983): 1397–1403.
11. Sabshin, ibid.
12. M. Sabshin. "Report of the medical director." *Am J Psychiatry* 143 (1986): 1342–6.
13. S. Frazier. "Report of the secretary, summary of meetings of Board of Trustees, May 1983–1984." *Am J Psychiatry* 141 (1984): 1314–21.
14. M. Sabshin. "Report of the medical director." *Am J Psychiatry* 142 (1985): 1243–6; and Sabshin, 1986, ibid.
15. See Sabshin's annual reports, 1982 and 1985.
16. M. Sabshin. *Changing American Psychiatry* (Washington, DC: American Psychiatric Publishing, 2008): 78.
17. Sabshin, 1983 report of medical director, ibid.
18. Sabshin, 1982 report of medical director, ibid.
19. Sabshin, 1985 report of medical director, ibid.
20. M. Sabshin. "Report of the medical director." *Am J Psychiatry,* 147 (1990): 1411–17.
21. Sabshin, 1985 report of medical director, ibid.
22. E. Benedek. "Report of the secretary: summary of actions of the Board of Trustees, May 1986–May 1987." *Am J Psychiatry* 144 (1987): 1381–8.
23. Benedek, ibid.
24. M. Sabshin. "Report of the medical director." *Am J Psychiatry* 144 (1987): 1390–3.
25. M. Sabshin. "Report of the medical director." *Am J Psychiatry* 149 (1992): 1434–44.
26. Sabshin, ibid.

27. Benedek, ibid.
28. M. Sabshin. "Report of the medical director." *Am J Psychiatry* 150 (1993): 1591–601.
29. M. Sabshin. "Report of the medical director." *Am J Psychiatry* 151 (1994): 1546–56.
30. See Sabshin's annual reports for 1992 and 1993, ibid.
31. M. Sabshin. "Report of the medical director." *Am J Psychiatry* 148 (1991): 1444–50.
32. American Psychiatric Association, Annual Report, 2005.
33. Ibid.
34. M. Riba. "Presidential Address." *Am J Psychiatry* 162 (2005): 2040–4.
35. American Psychiatry Association, Annual Report, 2006.
36. American Psychiatric Association, Annual Report, 2011.
37. J. Borenstein. "The council on communications." *Am J Psychiatry* 167 (2010): 228.
38. American Psychiatric Association, Annual Report, 2012.
39. A cited by P. Breggin, *Toxic Psychiatry* (New York: St. Martin's Press, 1991): 354.
40. G. Pollock. "Report of the treasurer." *Am J Psychiatry* 138 (1981): 1415–17.
41. J. Ronson. "Bipolar kids: victims of the 'madness industry'?" *New Scientist*, June 8, 2011.
42. "Report of the Secretary." *Am J Psychiatry* 138 (1981): 1408.
43. J. Scully. "Advertising Revenues Helps APA Meet Its Objectives." *Psychiatric News,* April 16, 2004. Also see S. Vedantam. "Industry role in medical meeting decried." *Washington Post,* May 26, 2002.
44. Scully, ibid.
45. Sabshin. *Changing American Psychiatry,* ibid, 193.
46. Annual reports of the treasurer published in *Am J Psychiatry,* 1980–2011.
47. F. Gottlieb. "Report of the speaker." *Am J Psychiatry* 142 (1985): 1248.
48. "Report of the secretary." *Am J Psychiatry* 140 (1983): 1389.
49. Gottlieb, ibid.
50. *Psychiatric News,* April 3, 1987; as cited by P. Breggin, *Toxic Psychiatry* (New York: St. Martin's Press, 1991): 347.
51. E. Benedek. "Report of the secretary." *Am J Psychiatry* 144 (1987): 1381–8; and *Am J Psychiatry* 145 (1988): 1328–36.
52. M. Sabshin. "Report of the medical director." *Am J Psychiatry* 145 (1988): 1340–2.
53. N. Rao. "The council on medical education and career development." *Am J Psychiatry* 158 (2001): 342–3; J. Greden. "The council on research." *Am J Psychiatry* 158 (2001): 346–8; J. Scully. "Report of the medical director." *Am J Psychiatry* 161 (2004): 1735–6.
54. Sabshin, 1992 report of the medical director, ibid.
55. Sabshin, *Changing American Psychiatry,* ibid, 62.
56. Annual reports of the treasurer published in *Am J Psychiatry,* 1980–2011.
57. C. Bernstein. "Report of the treasurer." *Am J Psychiatry* 158 (2001): 1760–1. For percentages in 2006 and 2008, see APA report. "Pharmaceutical Revenue 2010," which provided the amounts received from pharmaceutical companies from 2006 to 2010.
58. Sabshin, *Changing American Psychiatry,* ibid, 195.
59. "Report of the secretary." *Am J Psychiatry* 138 (1981): 1408.
60. C. Bernstein. "Report of the treasurer." *Am J Psychiatry* 159 (2002): 1622.
61. L. Cosgrove. "Pharmaceutical Philanthropic Shell Games." *Psychiatric Times,* March 6, 2010.
62. Cosgrove, ibid.
63. F. Gottlieb. "Report of the speaker." *Am J Psychiatry* 142 (1985): 1246–9.
64. Gottlieb, ibid.

65. A cited by P. Breggin, *Toxic Psychiatry*, ibid, 346.

66. M. Dumont. "In bed together at the market: psychiatry and the pharmaceutical industry." *Am J Orthopsychiatry* 60 (1990): 484–5.

67. Letter from Loren Mosher to Rodrigo Munoz, APA president, December 4, 1998.

68. C. Goldberg. " 'No' to Drug Money." *Boston Globe*, May 7, 2007.

69. P. Appelbaum. "Throw Them out?" *Psychiatric News,* July 5, 2002.

70. P. Appelbaum. "Dr. Appelbaum Responds." *Psychiatric News*, August 16, 2002.

71. E. Silverman. "Grassley probes psychiatrists over ties to pharma." Pharmalot.com, July 11, 2008.

72. B. Carey. "Psychiatric Group Faces Scrutiny over Drug Industry Ties." *New York Times*, July 12, 2008.

73. J. Yan. "New Guidelines Govern APA's Relations with Industry." *Psychiatric News,* June 18, 2010.

74. K. Hausman. "Industry-supported Symposia to be Phased out." *Psychiatric News,* April 17, 2009.

75. A. Schatzberg, APA Presidential Address, *Am J Psychiatry* 167 (2010): 1161–5.

76. American Psychiatric Association. "Pharmaceutical Revenue 2011."

77. Sabshin, *Changing American Psychiatry,* ibid.

78. American Psychiatric Association's "Program" for its 2008 annual meeting in Washington, DC; see "Full disclosure index."

79. L. Cosgrove. "Financial ties between DSM-IV panel members and the pharmaceutical industry." *Psychother Psychosom* 75 (2006): 154–60.

80. L. Cosgrove. "A comparison of DSM_IV and DSM-5 panel members' financial associations with industry." *PLoS Med* 9 (2012): e1001190.

81. L. Cosgrove. "Conflicts of interest and disclosure in the American Psychiatric Association's clinical practice guidelines." *Psychother Psychosom* 78 (2009): 228–32.

82. G. Fava. "Should the drug industry work with key opinion leaders? No." *BMJ* 336 (2008): 1405.

4 The Etiology of Mental Illness Is Now Known

1. D. Light. "Institutional corruption of pharmaceuticals and the myth of safe and effective drugs." *J Law Med Ethics* 41 (2013): 590–600.

2. G. Klerman. "The Evolution of a Scientific Nosology," in *Schizophrenia: Science and Practice,* edited by J. Shershow (Cambridge, MA: Harvard University Press, 1978): 104–5.

3. R. Spitzer. "A re-analysis of the reliability of psychiatric diagnosis." *Br J Psychiatry* 125 (1974): 341–7.

4. As cited by S. Kirk, *The Selling of DSM* (Hawthorne, NY: Aldine de Gruyter, 1992): 124.

5. R. Spitzer. "DSM-III field trials:1. Initial interrater diagnostic reliability." *Am J Psychiatry* 136 (1979): 815–17.

6. R. Spitzer. "DSM-III. The major achievements and an overview." *Am J Psychiatry* 137 (1980): 151–64.

7. American Psychiatric Association. *Diagnostic and Statistical Manual of Mental Disorders,* third edition (Washington, DC: American Psychiatric Association, 1980): 5, 468.

8. G. Klerman. "A debate on DSM-III." *Am J Psychiatry* 141 (1984): 539–53.

9. D. Langsley. "Presidential address: today's teachers and tomorrow's psychiatrists." *Am J Psychiatry* 138 (1981): 1013–16.

10. Klerman, ibid.
11. M. Strand. "Where do classifications come from? The DSM-III, the transformation of American psychiatry, and the problem of origins in the sociology of knowledge." *Theor Soc* 40 (2011): 273–313.
12. American Psychiatric Association, *DSM III*, ibid, 467.
13. M. Rutter. "DSM-III: a step forward or back in terms of the classification of child psychiatric disorders?" *J Am Acad Child Psychiatry* 19 (1980): 371–94.
14. American Psychiatric Association, *DSM-III*, ibid, 467–8.
15. H. Decker. *The Making of DSM-III* (New York: Oxford University Press, 2013): 258.
16. Decker, ibid, 255.
17. Kirk, ibid, 129.
18. Kirk, ibid, 121–60.
19. Kirk, ibid, 158.
20. Rutter, ibid.
21. Kirk, ibid, 56–63.
22. Kirk, ibid, 143, 171–3.
23. A. Bassett. "DSM-III: use of the multiaxial system in clinical practice." *Can J Psychiatry* 36 (1991): 270–4.
24. P. Lieberman. "The reliability of psychiatric diagnosis in the emergency room." *Hosp Community Psychiatry* 36 (1985): 291–3.
25. J. Anthony. "Comparison of the lay diagnostic interview schedule and a standardized psychiatric diagnosis. Experience in eastern Baltimore." *Arch Gen Psychiatry* 42 (1985): 667–75.
26. American Psychiatric Association. *DSM-III,* 8.
27. Klerman, ibid.
28. J. Talbott. "Presidential address: our patients' future in a changing world." *Am J Psychiatry* 142 (1985): 1003–8.
29. Associated Press. "Researcher says treatment may be near for depression," October 12, 1981.
30. N. Andreasen. *The Broken Brain* (New York: Harper & Row, 1984): 29.
31. Andreasen, ibid, 30, 133.
32. J. Franklin. "The Mind-Fixers." *Baltimore Sun*, series published in July 1985.
33. C. Nadelson. "Response to the presidential address." *Am J Psychiatry* 142 (1985): 1009–14.
34. D. Regier. "The NIMH depression awareness, recognition, and treatment program." *Am J Psychiatry* 145 (1988): 1351–7.
35. D. Healy, *Let Them Eat Prozac* (New York: New York University Press, 2004): 9.
36. K. Vick. "Society Clings to Idea that Mental Illness is a Stigma." *St. Petersburg Times*, August 6, 1988.
37. J. Lacasse. "Serotonin and depression: a disconnect between the advertisements and the scientific literature." *PLoS Med* 2 (2005): 1211–16.
38. P. Weiden, *Breakthroughs in Antipsychotics Medications* (New York: W.W. Norton, 1999): 26, 29.
39. R. Harding. "Unlocking the brain's secrets." *Family Circle*, November 20, 2001, 62.
40. N. Stotland. "About depression in women." *Family Circle*, November 20, 2001, 65.
41. American Psychiatric Association. "Mental illness stigmas are receding," press release, May 4, 2005.
42. American Psychiatric Association. "Let's talk facts about depression," 2005.
43. B. Pescosolido. "A disease like any other?" *Am J Psychiatry* 167 (2010): 1321–30; see table 1, p. 1324.

44. M. Bowers. "Lumbar CSF 5-hydroxyindoleacetic acid and homovanillic acid in affective syndromes." *J Nerv Ment Disease* 158 (1974): 325–30.

45. R. Papeschi. "Homovanillic and 5-hydroxyindoleacetic acid in cerebrospinal fluid of depressed patients." *Arch Gen Psychiatry* 25 (1971): 354–8.

46. Bowers, ibid.

47. J. Mendels. "Brain biogenic amine depletion and mood." *Arch Gen Psychiatry* 30 (1974): 447–51.

48. J. Maas. "Pretreatment neurotransmitter metabolite levels and response to tricyclic antidepressant drugs." *Am J Psychiatry* 141 (1984): 1159–71.

49. S. Dubovsky. "Mood disorders," in *Textbook of Psychiatry*, edited by R. Hales, third edition (Washington, DC: American Psychiatric Press, 1999): 516.

50. Lacasse, ibid.

51. V. Krishnan. "Linking molecules to mood." *Am J Psychiatry* 167 (2010): 1305–20.

52. National Public Radio. "When it comes to depression, serotonin isn't the whole story." January 23, 2012.

53. M. Bowers. "Central dopamine turnover in schizophrenic syndromes." *Arch Gen Psychiatry* 31 (1974): 50–4. R. Post. "Cerebrospinal fluid amine metabolites in acute schizophrenia." *Arch Gen Psychiatry* 32 (1975): 1063–9.

54. T. Lee. "Binding of ^{3}H-neuroleptics and ^{3}H-apomorphine in schizophrenic brains." *Nature* 274 (1978): 897–900.

55. J. Kornhuber. "3H-spiperone binding sites in post-mortem brains from schizophrenic patients." *J Neural Transmission* 75 (1989): 1–10.

56. J. Kane. "Towards more effective antipsychotic treatment." *Br J Psychiatry* 165, suppl. 25 (1994): 22–31.

57. E. Nestler, *Molecular Neuropharmacology* (New York: McGraw Hill, 2001): 392.

58. A. Jucaite. "Dopaminergic hypothesis of schizophrenia," in *Emerging Therapies for Schizophrenia*, edited by J. Albert (Hoboken, NJ: John Wiley & Sons, 2012): 5–35.

59. Lacasse, ibid.

60. T. Insel. "Director's blog: mental illness defined as disruption in neural circuits." Published on nimh.nih.gov, August 12, 2011.

61. R. Pies. "Psychiatry's New Brain-mind and the Legend of the 'Chemical Imbalance.'" *Psychiatric Times*, July 11, 2011.

62. R. Pies. "Doctor, is my mood disorder due to a chemical imbalance?" Blog, August 4, 2011, published on PsychCentral.com.

63. C. Dean. "Psychiatry revisited." *Minnesota Medicine*, March 2008, 40–4.

64. M. England. "Response to the presidential address: one system of equal health care for all." *Am J Psychiatry* 152 (1995): 1252–3.

65. N. Andreasen, editorial. "The validation of psychiatric diagnosis." *Am J Psychiatry* 152 (1995): 161–2.

66. C. Robinowitz. " Response to the presidential address." *Am J Psychiatry* 164 (2007): 1510–13.

67. J. Phillips. "The six most essential questions in psychiatric diagnosis: a pluralogue part 2." *Philosophy Ethics Humanities Med* 7 (2012): 8.

68. J. Phillips. "The six most essential questions in psychiatric diagnosis: a pluralogue part 1." *Philosophy Ethics Humanities Med* 7 (2012): 3.

69. T. Insel. "Director's blog: transforming diagnosis." Published on nimh.nih.gov, April 29, 2013.

70. N. Andreasen. "DSM and the death of phenomenology." *Schizophrenia Bull* 33 (2007): 108–12.

71. Phillips, ibid, part 1.

72. Phillips, ibid, part 2.
73. Phillips, ibid, part 2.
74. S. Ghaemi. "Requiem for DSM." *Psychiatric Times*, July 17, 2013.

5   Psychiatry's New Drugs

1. E. Turner. "Publication bias, with a focus on psychiatry." *CNS Drugs* 27 (2013): 457–68.
2. Medco. "America's State of Mind," November 2011.
3. A. Tone, *The Age of Anxiety.* (New York: Basic Books, 2009): 153.
4. Tone, ibid, 176.
5. M. Smith, *Small Comfort* (New York: Praeger, 1985): 78.
6. Tone, ibid, 212.
7. G. Klerman. "Overview of the cross-national collaborative panic study." *Arch Gen Psychiatry* 45 (1988): 407–12.
8. Klerman, ibid.
9. R. Noyes. "Alprazolam in panic disorder and agoraphobia: II. *Arch Gen psychiatry* 45 (1988): 423–8.
10. J. Pecknold. "Alprazolam in panic disorder and agoraphobia." *Arch Gen Psychiatry* 45 (1988): 429–36.
11. Klerman, ibid.
12. Klerman, Ibid; see editorial note on page 212 stating that Freedman served on a committee that served as "consultants to the Upjohn Company."
13. Klerman, Ibid.
14. Noyes, ibid.
15. Pecknold, ibid.
16. Klerman, ibid.
17. P. Breggin, *Toxic Psychiatry* (New York: St. Martin's Press, 1991): 344–53.
18. F. Pollner. "Don't overlook panic disorder." *Medical World News,* October 1, 1991.
19. M. Sabshin. "To Aid Understanding of Mental Disorders." *New York Times*, March 10, 1992.
20. Sabshin, ibid.
21. J. Randal. "In a Panic?" *St. Louis Post-Dispatch*, October 7, 1990.
22. T. Plyler. "Treatment of a Panic Disorder Starts with Recognizing Problem." *The Union Leader*, February 23, 1992.
23. J. Cornwell, *The Power to Harm* (New York: Viking, 1996): 147–8.
24. D. Graham. "Sponsor's ADR submission on fluoxetine dated July 17, 1990," FDA document, September 1990.
25. I. Kirsch. "The emperor's new drugs: an analysis of antidepressant medication data submitted to the U.S. Food and Drug Administration." *Prevention & Treatment* 5, July 15, 2002.
26. FDA. "Review and Evaluation of Efficacy Data," March 28, 1985. As cited by P. Breggin, *Talking Back to Prozac* (New York: St. Martin's Press, 1994): 46.
27. Kirsch, ibid.
28. J Bremner. "Fluoxetine in depressed patients." *J Clin Psychiatry* 45 (1984): 414–19.
29. Breggin, *Talking Back to Prozac,* ibid, 45.
30. J. Cohn. "A comparison of fluoxetine, imipramine, and placebo in patients with major depressive disorder." *J Clin Psychiatry* 46 (1985): 26–31.
31. Breggin, ibid, 43.

32. W. Byerley. "Fluoxetine, a selective serotonin uptake inhibitor, for the treatment of outpatients with major depression." *J. Clin Psychopharmacol* 8 (1988): 112–15.

33. Breggin, ibid, 44.

34. J. Feighner. "A double-blind comparison of fluoxetine, imipramine, and placebo in outpatients with major depression." *Int Clin Psychopharmacol* 4 (1989): 127–34.

35. Breggin, ibid, 42.

36. Breggin, ibid, 46.

37. S. Levine. "A comparative trial of a new antidepressant, fluoxetine." *Br J Psychiatry* 150 (1987): 653–5.

38. R. Pary. "Fluoxetine: prescribing guidelines for the newest antidepressant." *South Med J* 82 (1989): 1005–9.

39. B. Runck. "NIMH to launch major campaign on recognition and treatment of depression." *Hosp Community Psychiatry* 37 (1986): 779–80.

40. D. Regier. "The NIMH depression awareness, recognition, and treatment program." *Am J Psychiatry* 145 (1988): 1351–7.

41. Regier, ibid.

42. Runck, ibid.

43. Regier, ibid.

44. D. Healy, *Let Them Eat Prozac* (New York: New York University Press, 2004): 9.

45. G. Cowley. "Prozac: a Breakthrough Drug for Depression." *Newsweek*, March 26, 1990.

46. N. Angier. "New Antidepressant is Acclaimed, but Not Perfect." *New York Times,* March 29, 1990.

47. P. Elmert-DeWitt. "Depression: The Growing Role of Drug Therapies." *Time*, July 6, 1992.

48. As cited in a legal complaint by plaintiff, *Plumlee v. Pfizer*, filed January 30, 2013; page 10 of complaint.

49. FDA, Psychopharmacological Drugs Advisory Committee, November 19, 1990, p. 101.

50. Ibid, 90.

51. Ibid, 69.

52. Ibid, 99.

53. FDA's statistical review of sertraline; citation accessed on February 16, 2014 at http://1boringoldman.com/index.php/2013/02/17/zoloft-the-approval-i/.

54. L. Fabre. "Sertraline safety and efficacy in major depression." *Biol Psychiatry* 38 (1995): 592–602.

55. D. Casey. "Striking a balance between safety and efficacy: experience with the SSRI sertraline." *Int Clin Psychopharmacol* 9 suppl. 3 (1994): 5–12.

56. See disclosure for D. Casey on Medscape, for a CME program, "Optimizing treatment for patients with schizophrenia: targeting positive outcomes." That CME program was based on a presentation that Casey gave on May 20, 2003, at the APA's 156th annual meeting.

57. E. Turner. "Selective publication of antidepressant trials and its influence on apparent efficacy." *New Engl J Med* 358 (2008): 252–60.

58. L. McHenry. "Of sophists and spin-doctors." *Mens Sana Monogr* 8 (2010): 129–45.

59. I. Kirsch. "Antidepressants and placebos: secrets, revelations, and unanswered questions." *Prevention & Treatment* 5, July 15, 2002.

60. E. Turner. "Selective publication of antidepressant trials and its influence on apparent efficacy." *New Engl J Med* 358 (2008): 252–60.

61. I. Kirsch. "Initial severity and antidepressant benefits." *PLoS Med* 5 (2008): 260–8.

62. G. Parker. "Antidepressants on trial." *Br J Psychiatry* 194 (2009): 1–3.

63. CBS News, *Sixty Minutes*: "The placebo phenomenon." February 19, 2012.

64. FDA reviews of New Drug Application for risperidone. Reviews of risperidone data include the following written commentaries: reviews by Andrew Mosholder, May 11, 1993 and November 7, 1993; David Hoberman, April 20, 1993; Thomas Laughren, December 20, 1993; and Paul Leber, December 21, 1993.

65. Approval letter from Robert Temple to Janssen Research Foundation, December 29, 1993.

66. S. Marder. "Risperidone in the treatment of schizophrenia." *Am J Psychiatry* 151 (1994): 825–35. Also L. Cohen. "Risperidone." *Pharmacotherapy* 14 (1994): 253–65.

67. S. Marder. "The effects of risperidone on the five dimensions of schizophrenia derived by factor analysis." *J Clin Psychiatry* 58 (1997): 538–46.

68. M. Freudenheim. "Seeking Safer Treatments for Schizophrenia." *New York Times*, January 15, 1992.

69. FDA review of New Drug Application for olanzapine. FDA reviews of olanzapine data include the following written commentaries: reviews by Thomas Laughren on September 27, 1996; by Paul Andreason on July 29 and September 26, 1996; and by Paul Leber on August 18 and August 30, 1996.

70. C. Beasley. "Efficacy of olanzapine." *J Clin Psychiatry* 58, suppl. (1997): 7–12; G. Tollefson. "Depressive signs and symptoms in schizophrenia." *Arch Gen Psychiatry* 55 (1998): 250–8; S. Hamilton. "Olanzapine versus placebo and haloperidol." *Neuropsychopharmacology* 18 (1998): 41–9.

71. T. Burton. "Psychosis Drug from Eli Lilly Racks Up Gains." *Wall Street Journal,* April 14, 1998.

72. Associated Press. "A New Drug for Schizophrenia Wins Approval from the FDA." *New York Times*, October 2, 1996.

73. FDA review of New Drug Application for quetiapine. FDA reviews of quetiapine data include the following written commentaries: reviews by Andrew Mosholder on June 13 and August 19, 1997; Thomas Laughren on August 21, 1997; and Paul Leber on September 24, 1997.

74. G. Spielmans. "From evidence-based medicine to marketing-based medicine." *J Bioethic Inq* 7 (2010): 13–29.

75. L. Arvanitis. "Multiple doses of 'Seroquel' (quetiapine) in patients with acute exacerbation of schizophrenia." *Biol Psychiatry* 42 (1997): 233–46.

76. Spielmans, ibid. Tumas's e-mail, which is reprinted in this article, is dated December 6, 1999.

77. J. Olson. "U Doctor Scrutinized over Drug Research." *St. Paul Pioneer Press,* March 18, 2009.

78. S. Roan. "Lives Recovered." *Los Angeles Times*, January 30, 1996. Quote is from the headline.

79. "Schizophrenia Project Launched." *Psychiatric News*, July 18, 1997.

80. P. Huston. "Redundancy, disaggregation, and the integrity of medical research." *Lancet* 347 (1996): 1024–6.

81. R. Horton. "Prizes, publications, and promotion." *Lancet* 348 (1996): 1398.

82. J. Geddes. "Atypical antipsychotics in the treatment of schizophrenia." *BMJ* 321 (2000): 1371–6.

83. R. Rosenheck. "Effectiveness and cost of olanzapine and haloperidol in the treatment of schizophrenia." *JAMA* 290 (2003): 2693–702.

84. J. Lieberman. "Effectiveness of antipsychotic drugs in patients with chronic schizophrenia." *New Engl J Med* 353 (2005): 1209–33.

85. L. Davies. "Cost-effectiveness of first- v. second-generation antipsychotic drugs." *Br J Psychiatry* 191 (2007): 14–22.

86. J. Lieberman. "Comparative effectiveness of antipsychotic drugs." *Arch Gen Psychiatry* 63 (206): 1069–72.

87. S. Vedantam. "In Antipsychotics, Newer Isn't Better Drug Find Shocks Researchers." *Washington Post*, October 3, 2006.

88. P. Tyrer. "The spurious advance of antipsychotic drug therapy." *Lancet* 373 (2009): 4–5.

6 Expanding the Market

1. J. Maracek. "Disappearances, silences and anxious rhetoric: gender in abnormal psychology textbooks." *J Theor Philos Psychol* 13 (1993): 114–23.

2. J. Biederman. "Attention-deficit/hyperactivity disorder: a life-span perspective." *J Clin Psychiatry* 59, suppl. 7 (1998): 4–16.

3. M. Strand. "Where do classifications come from? The DSM-III, the transformation of American psychiatry, and the problem of origins in the sociology of knowledge." *Theor Soc* 40 (2011): 273–313.

4. *Diagnostic and Statistical Manual of Mental Disorders*, fourth edition (Washington DC: American Psychiatric Association, 1994): xxii.

5. L. Cosgrove. "Financial ties between DSM-IV panel members and the pharmaceutical industry." *Psychother Psychosom* 75 (2006): 154–60.

6. R. Kessler. "Prevalence, severity, and comorbidity of 12-month DSM-IV disorders in the National Comorbidity Survey Replication." *Arch Gen Psychiatry* 62 (2005): 617–27.

7. Centers for Disease Control and Prevention, National Health and Nutrition Examination Survey, 2012.

8. R. Mayes. *Medicating Children* (Cambridge, MA: Harvard University Press, 2009): 46.

9. G. Jackson. "Postmodern psychiatry," unpublished paper, September 2, 2002.

10. Mayes, ibid, 61.

11. *Diagnostic and Statistical Manual of Mental Disorders*, third edition (Washington DC: American Psychiatric Association, 1980): 42.

12. *DSM-IV,* ibid, 78–85.

13. B. Lahey. "DSM-IV field trials for attention deficit hyperactivity disorder in children and adolescents." *Am J Psychiatry* 151 (1994): 1673–85.

14. A PubMed search of "Biederman ADHD" produces a record of 604 articles from 1984 to 2013, with Biederman, J. listed as an author or coauthor. Some of these articles are on other pediatric disorders, such as conduct disorder or bipolar disorder, and thus not all are focused on ADHD.

15. S. Faraone. "Is attention deficit hyperactivity disorder familial?" *Harv Rev Psychiatry* 1 (1994): 271–87.

16. Lahey, ibid.

17. Biederman, ibid.

18. T. Wilens. "Attention-deficit hyperactivity disorder and comorbid substance use disorder in adults." *Psychiatr Serv* 46 (1995): 761–3.

19. T. Spencer. "Pharmacotherapy of attention-deficit hyperactivity disorder across the life cycle." *J Am Acad Child Adolesc Psychiatry* 35 (1996): 409–32.

20. Wilens, ibid.

21. G. Kolata. "Boom in Ritalin Sales Raises Ethical Issues." *New York Times,* May 15, 1996.

22. A. Schwarz. "The Selling of Attention Deficit Disorder." *New York Times,* December 14, 2013.

23. See disclosure in A. Kotte. "Autistic traits in children with and without ADHD." *Pediatrics* 132 (2013): e612–22.

24. See Centers for Disease Control. "Summary Health Statistics for U.S. Children: National Health Survey, 2012" for report that 10 percent of children 3–17 were diagnosed with ADHD; see Express Scripts. "Turning Attention To ADHD," March 2014, for report that 7.75 percent of boys ages 4 to 18, and 3.5 percent of girls ages 4 to 18, took an ADHD medication in 2012.

25. M. Smith. *Small Comfort* (New York: Praeger, 1985): 32.

26. IMS Health. "Top Therapeutic Classes by U.S. Dispensed Prescriptions," 2012.

27. A. Horwitz. *The Loss of Sadness* (Oxford, NY: Oxford University Press, 2007): 53–104.

28. Ibid, 59.

29. Ibid, 62.

30. Ibid, 67.

31. Ibid, 75–82.

32. Ibid, 74.

33. Ibid. 87.

34. *DSM-III,* ibid, 211–15.

35. Horwitz, ibid, 103.

36. C. Silverman. *The Epidemiology of Depression* (Baltimore, MD: Johns Hopkins Press, 1968): 45–7.

37. D. Regier. "The de facto U.S. mental and addictive disorders service system." *Arch Gen Psychiatry* 50 (1993): 85–94.

38. R. Kessler. "Lifetime and 12-month prevalence of DSM-III-R psychiatric disorders in the United States. Results from the National Comorbidity Survey." *Arch Gen Psychiatry* 51 (1994): 8–19.

39. R. Hirschfeld. "The National Depressive and Manic-Depressive Association consensus statement on undertreatment of depression." *JAMA* 277 (1997): 333–40.

40. *DSM-IV,* ibid, 317–28, 623.

41. *DSM-III,* ibid, 225–39.

42. Kessler, ibid.

43. Scientific Therapeutics Information. "Paxil Advisory Board Meeting: A proposal for one psychiatrist advisory board meeting." October 1, 1993. Accessed through Project on Government Oversight.

44. J Ballenger. "Double-blind, fixed-dose, placebo-controlled study of paroxetine in the treatment of panic disorder." *Am J Psychiatry* 155 (1998): 36–42.

45. D. Sheehan. "The role of SSRIs in panic disorder." *J Clin Psychiatry* 57, suppl. 10 (1996): 51–8.

46. D. Sheehan. "Current concepts in the treatment of panic disorder." *J Clin Psychiatry* 60, suppl. 18 (1999): 16–21.

47. M. Liebowitz. "Social phobia. Review of a neglected anxiety disorder." *Arch Gen Psychiatry* 42 (1985): 729–36.

48. See *DSM-III,* ibid, 228, and *DSM-IV,* ibid, 243.

49. M. Pollack. "Introduction. New frontiers in the management of social anxiety disorder: diagnosis, treatment, and clinical course." *J Clin Psychiatry* 60, suppl. 9 (1999): 3–8.

50. For financial disclosure, see M. Liebowitz. "Anxiety symptoms and treating depression." *Medscape*, March 25, 2004.

51. M. Stein and M. Liebowitz. "Paroxetine treatment of generalized social phobia (social anxiety disorders): a randomized controlled trial." *JAMA* 280 (1998): 708–13.

52. M. Liebowitz. "Update on the diagnosis and treatment of social anxiety disorder." *J Clin Psychiatry* 60, suppl. 18 (1999): 22–6.

53. See Davidson's disclosure in Hypericum Depression Trial Study Group. "Effect of *Hypericum perforatum* (St John's Wort) in major depressive disorder." *JAMA* 287 (2002): 1807–14.

54. J. Davidson. "Introduction: new strategies for the treatment of posttraumatic stress disorder." *J Clin Psychiatry* 61, suppl. 7 (2000): 3–4. Also see R. Hidalgo and J. Davidson. "Postraumatic stress disorder: epidemiology and health-related considerations." *J Clin Psychiatry* 61, suppl. 7 (2000): 5–13.

55. J. Davidson. "Treatment of posttraumatic stress disorder: the impact of paroxetine." *Psychopharmacol Bull* 37, suppl. 1 (2003): 76–88.

56. J. Ballenger. "Consensus statement on generalized anxiety disorder from the International Consensus Group on Depression and Anxiety." *J Clin Psychiatry* 62, suppl. 11 (2001): 53–8.

57. Fax from Oliver Dennis of The Medicine Group to Scott Sproul, January 21, 1999. Document produced by GSK in In Re Paxil, C.P.Ct.Pa, accessed through Drug Industry Document Archive, University of California San Francisco. (DIDA).

58. Update, Paxil Publication Plan. Produced by GSK in *Steele v. GSK*. Accessed through DIDA.

59. C. Nemeroff. "Advancing the treatment of mood and anxiety disorders." *Psychopharmacol Bull* 37, suppl. 1 (2003): 6–7.

60. STI Cover Page, First draft of Paroxetine Treatment of Mood Disorders in Women, March 24, 2003. Produced by GSK in *Crooks v. SmithKline*. Accessed through DIDA.

61. Nemeroff, ibid.

62. Psychnet, PAXIL Clinicians Speaker Council. Document produced by GSK in *Cunningham vs. GSK*. Accessed through DIDA.

63. PsychNet Program Workbook, March 10–12, 2000, Naples Florida. Documents accessed through DIDA.

64. Deirdre Zuccarello e-mail to Dr. Sheehan, September 13, 2000. Document accessed through DIDA.

65. Letter from Senator Charles Grassley to James Wagner, Emory University President, September 16, 2008. Accessed through finance.senate.gov.

66. R. Kessler. "Mood disorders in children and adolescents." *Biol Psychiatry* 49 (2001): 1002–14.

67. Keller, ibid.

68. K. Merikangas. "Life prevalence of mental disorders in U.S. adolescents: Results from the National Comorbidity Study-Adolescent Supplement." *J Am Acad Child Adolesc Psychiatry* 49 (2010): 980–9.

69. Kessler, ibid.

70. M. Keller. "Depression in children and adolescents." *J Affect Disorders* 21 (1991): 163–71.

71. Letter from Martin Keller to Catherine Sohn, SmithKline Beecham, March 19, 1993. Produced by GSK in "In Re Paxil, C.P.Ct.Pa," accessed through DIDA.

72. B. Strauch. "Use of Antidepression Medicine for Young Patients Has Soared." *New York Times*, August 10, 1997.

73. G. Emslie. "A double-blind, randomized, placebo-controlled trial of fluoxetine in children and adolescents with depression." *Arch Gen Psychiatry* 54 (1997): 1031–7.
74. G. Emslie. "Mood disorders in children and adolescents." *Biol Psychiatry* 49 (2001): 1082–90. Also see "Breakthroughs in the treatment of child and adolescent depression and anxiety," newsletter produced by GSK in "Smith v. GSK," accessed through DIDA.
75. M. Goozner. "SSRI use in children: an industry-biased record." Center for Science in the Publish Interest, February 2004.
76. A. Bass. "Drug Companies Enrich Brown Professor." *Boston Globe*, October 4, 1999.
77. "Payments to physicians." *Congressional Record* 154, Number 142 (September 9, 2008): s8165–9.
78. L. Kowalczyk. "US cites Boston Psychiatrist in Case vs. Drug Firm." *Boston Globe*, March 6, 2009.
79. T. Delate. "Trends in the use of antidepressants in a national sample of commercially insured pediatric patients, 1998 to 2002." *Psychiatr Serv* 55 (2004): 387–91.
80. T. Laughren, Memorandum. "Background comments for Feb. 2 meeting of psychopharmacological drugs advisory committee," January 4, 2004. Accessed at fda.gov.
81. Editorial. "Depressing research." *Lancet* 363 (2004): 1335.
82. J. Leo. "The SSRI trials in children." *Ethical Human Psychology and Psychiatry* 8 (2006): 29–41.
83. E. Jane Garland. "Facing the evidence: antidepressant treatment in children and adolescents." *CMAJ* 170 (February 17, 2004): 489–91. Also see Leo, ibid.
84. Leo, ibid.
85. *United States vs. GlaxoSmithKline*, Complaint, October 26, 2011.
86. SmithKline document, prepared by CMAt-Neurosciences, marked "confidential": "Seroxat/Paxil adolescent depression: position piece on the phase III clinical studies." October 1998.
87. J. Jureidini. "Clinical trials and drug promotion." *Int J Risk Saf Medicine* 20 (2008): 73–81.
88. *United States vs. GlaxoSmithKline*, Complaint, October 26, 2011.
89. Jureidini, ibid.
90. M. Keller. "Efficacy of paroxetine in the treatment of adolescent major depression." *J Am Acad Child Adolesc Psychiatry* 40 (2001): 762–72.
91. K. Wagner. "Efficacy of sertraline in the treatment of children and adolescents with major depressive disorder." *JAMA* 290 (2003): 1033–41. Also see Leo, ibid.
92. Leo, ibid. Also see Wagner, ibid.
93. Garland, ibid.
94. D. Healy. "One flew over the conflict of interest net." *World Psychiatry* 6 (2007): 26–7.
95. C. Silverman. *The Epidemiology of Depression* (Baltimore, MD: Johns Hopkins University Press, 1968): 35, 139. Given that in the studies of the 1940s and 1950s, fewer than 25 percent of the manic-depressive patients were of the manic or cycling types, at most 25 percent of the manic-depressive group in the hospital would have been of the bipolar type.
96. *DSM-III*, ibid, 217.
97. H. Akiskal. "Cyclothymic disorder: validating criteria for inclusion in the bipolar affective group." *Am J Psychiatry* 134 (1977): 1227–33.
98. *DSM-IV*, ibid, 317–8; 353, 360, 364.

99. H. Akiskal. "The prevalent clinical spectrum of bipolar disorders: beyond DSM-IV." *J Clin Psychopharmacol* 16, 2 suppl. 1 (1996): 4S–14S.

100. H Askiskal. "The emergence of the bipolar spectrum." *Psychopharmacol Bull* 40 (2007): 99–115.

101. H. Akiskal. "Re-evaluating the prevalence of and diagnostic composition within the broad clinical spectrum of bipolar disorders." *J Affect Disorders* 59, suppl. 1 (2000): S5–S30.

102. L. Judd. "The prevalence and disability of bipolar spectrum disorders in the US population." *J Affect Disorders* 73 (2003): 123–31.

103. H Akiskal. "The emergence of the bipolar spectrum." *Psychopharmacol Bull* 40 (2007): 99–115.

104. See disclosure statement for Akiskal in G. Parker. "Issues for DSM-5: Whither melancholia? The case for its classification as a distinct mood disorder." *Am J Psychiatry* 167 (2010): 745–7.

105. G. Spielmans. "From evidence-based medicine to marketing-based medicine: Bioethic Inquiry." *J Bioethic Inq* 7 (2010): 13–29.

106. Spielmans, ibid.

107. Spielmans, ibid.

108. N. Ghaemi. "The failure to know what isn't known: negative publication bias with lamotrigine and a glimpse inside peer review." *Evid Based Ment Health* 12 (2009): 65–8.

109. Ghaemi, ibid.

110. Abramson J. Expert report of John Abramson, MD: http://dida.library.ucsf.edu/pdf/oxx18v10. Neurontin Marketing and Sales Practices Litigation, MDL #1629, Docket #04–10981. 2008 August 11, p. 52–3.

111. Abramson, ibid, 56.

112. Abramson, ibid, 100.

113. Abramson, ibid, 101.

114. IMS Health. "2009 top therapeutic classes by U.S. sales."

115. L. Cosgrove. "A comparison of DSM-IV and DSM-5 panel members' financial associations with industry." *PLoS Med* 9 (2012): e1001190.

116. American Psychiatric Association. "Highlights of Changes from DSM-IV-TR to DSM-5," 2013. Also see B. Stetka. "A guide to DSM-5" *Medscape*, May 21, 2013.

117. L. Cosgrove. "Tripartite conflicts of interest and high stakes patent extensions in the DSM-5." *Psychother Psychosom* 83 (2014): 106–13.

118. Cosgrove, ibid.

119. T. Insel. "Director's blog: transforming diagnosis." Published on nimh.nih.gov, April 29, 2013.

120. D. Regier. "DSM-5 field trials in the United States and Canada. Part II: test-retest reliability of selected categorical diagnoses." *Am J Psychiatry* 170 (2013): 59–70.

121. S. Vanheule. *Diagnosis and the DSM: A Critical Review* (New York, NY, Palgrave Macmillan, 2014): 40.

122. J. Ronson. "Bipolar kids: victims of the 'madness industry'?" *New Scientist* 2815, June 8, 2011: 44–7.

7   Protecting the Market

1. P. Thagard. "The moral psychology of conflicts of interest." *J Applied Philosophy* 24 (2007): 367–80.

2. J. Richters. "NIMH Collaborative Multisite Multimodal Treatment Study of Children with ADHD." *J Am Acad Child Adolesc Psychiatry* 34 (1995): 987–1000.

3. The MTA Cooperative Group. "A 14-month randomized clinical trial of treatment strategies for attention-deficit/hyperactivity disorder." *Arch Gen Psychiatry* 56 (1999): 1073–86.

4. P. Jensen. "3-year follow-up of the NIMH MTA Study." *J Am Acad Child Adolesc Psychiatry* 46 (2007): 989–1002.

5. B. Molina. "Delinquent behavior and emerging substance use in the MTA at 36 months." *J Am Acad Child Adolesc Psychiatry* 46 (2007): 1028–40.

6. J. Swanson. "Evidence, interpretation and qualification from multiple reports of long-term outcomes in the multimodal treatment study of children with ADHD Part II." *J Attention Disorders* 12 (2008): 15–43.

7. B. Molina. "MTA at 8 years." *J Am Acad Child Adolesc Psychiatry* 48 (2009): 484–500.

8. Swanson, ibid.

9. Jensen, ibid.

10. NIMH press release. "Improvement Following ADHD treatment Sustained in Most Children," July 20, 2007.

11. Molina. MTA at 8 years, ibid.

12. Western Australian Department of Health. "Raine ADHD study: Long-term outcomes associated with stimulant medication in the treatment of ADHD children," 2009. Accessed at: www.health.wa.gov.au/publications/documents/MICADHD_Raine_ADHD_study_report_022010.pdf

13. J. Currie. "Do stimulant medications improve educational and behavioral outcomes for children with ADHD?" NBER working paper 19105, June 2013.

14. American Academy of Child and Adolescent Psychiatry and American Psychiatric Association. "ADHD Parents Medication Guide." Revised July 2013.

15. M. Posternak. "A reevaluation of the exclusion criteria used in antidepressant efficacy trials." *Am J Psychiatry* 159 (2002): 191–200.

16. J. Rush. "One-year clinical outcomes of depressed public sector outpatients." *Biol Psychiatry* 56 (2004): 46–53.

17. Rush, ibid.

18. Hypericum depression trial study group. "Effect of Hypericum perforatum (St. John's Wort) in major depressive disorder." *JAMA* 287 (2002): 1807–14.

19. M. Babyak. "Exercise treatment for major depression." *Psychosom Med* 62 (2000): 633–38.

20. NIMH. "Questions and answers about the NIMH Sequenced Treatment Alternatives to Relieve Depression (STAR*D) Study—Background." Press release, January 2006.

21. J. Rush. "Sequenced treatment alternatives to relieve depression (STAR*D): rationale and design." *Controlled Clinical Trials* 25 (2004): 119–42.

22. STAR*D protocol, 48.

23. M. Trivedi. "Evaluation of outcomes with citalopram for depression using measurement-based care in STAR*D." *Am J Psychiatry* 163 (2006): 28–40.

24. A. Rush. "Buproprion-SR, sertraline, or venlafaxine-XR after failures of SSRIs for depression." *NEJM* 354 (2006): 1231–42. Also see M. Trivedi. "Medication augmentation after the failure of SSRIs for depression." *NEJM* 354 (2006): 1243–52.

25. A. Rush. "Acute and longer-term outcomes in depressed outpatients requiring one or several treatment steps: A STAR*D report." *Am J Psychiatry* 163 (2006): 1905–17.

26. NIMH press release. "Questions and answers about the NIMH Sequenced Treatment Alternatives to Relieve Depression (STAR★D) Study—all medication levels." November 2006.
27. NIMH press release. "New strategies help depressed patients become symptom-free." March 23, 2006.
28. L. Menand. "Head Case." *The New Yorker*, March 1, 2010.
29. E. Pigott. "Adding Fiction to Fiction," blog post on madinamerica.com, April 10, 2011.
30. H. Pigott. "Efficacy and effectiveness of antidepressants." *Psychother Psychosom* 79 (2010): 267–79. Also see H. Pigott. "STAR★D: A tale and trail of bias." *Ethical Human Psychology and Psychiatry* 13 (2011): 6–28.
31. D. Brauser. "Broad review of FDA trials suggests antidepressants only marginally better than placebo." *Medscape Medical News*, August 24, 2010.
32. A. Rush. "Acute and longer-term outcomes in depressed outpatients requiring one or several treatment steps: A STAR★D report." *Am J Psychiatry* 163 (2006): 1905–17.
33. L. Pratt. "Antidepressant use in persons age 12 and over: United States 2005–2008." National Center for Health Statistics Data Brief # 76, October 2011.
34. J. Jureidini. "Efficacy and safety of antidepressants for children and adolescents." *Br Med J* 328 (2004): 879–83.
35. C. Whittington. "Selective serotonin reuptake inhibitors in childhood depression." *Lancet* 363 (2004): 1341–5.
36. TADS team. "Fluoxetine, cognitive-behavioral therapy, and their combination for adolescents with depression." *JAMA* 292 (2004): 807–20.
37. G. Emslie. "Treatment for adolescents with depression study (TADS): Safety results." *J Am Acad Child Adolesc Psychiatry* 45 (2006): 1440–55.
38. TADS team. "The treatment for adolescents with depression study (TADS): Long-term effectiveness and safety outcomes." *Arch Gen Psychiatry* 64 (2007): 1132–44.
39. B. Kennard. "Assessment of safety and long-term outcomes of initial treatment with placebo in TADS." *Am J Psychiatry* 166 (2009): 337–44.
40. B. Vitiello. "Suicidal events in the treatment for adolescents with depression study (TADS)." *J Clin Psychiatry* 70 (2009): 741–7.
41. See disclosure statements for the TADS studies cited above.

8   The End Product: Clinical Practice Guidelines

1. D. Antonuccio. "Psychology in the prescription era: building a firewall between marketing and science." *American Psychologist*, December 2003, 1028–43.
2. G. Guyatt. *User's Guides to the Medical Literature: A Manual for Evidence-Based Practice* (Chicago: American Medical Association, 2002): xiv.
3. K. Dickersin. "Evidence based medicine: Increasing, not dictating, choice." *Br Med Journal* 334 (2007): s10. Also see V. Montori. "Progress in evidence-based medicine." *JAMA: The Journal of the American Medical Association* 200 (2008): 1814–16.
4. IOM (Institute of Medicine). *Clinical Practice Guidelines We Can Trust* (Washington, DC: The National Academies Press, 2011): 4.
5. J. Hitt. "The Year in Ideas: A. to Z.; Evidence-based Medicine." *The New York Times*, December 9, 2001.
6. Dickersin, ibid.
7. T. Shaneyfelt. "Reassessment of clinical practice guidelines." *JAMA* 301 (2009): 868–9.

8. N. Choudhry. "Relationships between authors of clinical practice guidelines and the pharmaceutical industry." *JAMA* 287 (2002): 612–17.
9. S. Norris. "Conflict of interest in clinical practice guideline development." *PLoS ONE* 6 (2011): e25153. A. Shaughnessy. "What happened to the valid POEMs? A survey of review articles on the treatment of type 2 diabetes." *Br Med J* 327 (2003): 266. P. Rothwell. "External validity of randomised controlled trials: 'To whom do the results of this trial apply?'" *Lancet* 365 (2005): 82–93. M. Krahn. "The next step in guideline development: Incorporating patient preferences." *JAMA* 300 (2008): 436–8. C. Chong. "How well do guidelines incorporate evidence on patient preferences?" *J Gen Intern Med* 24 (2009): 977–82.
10. Norris, ibid.
11. IOM, ibid, 2.
12. IOM, ibid.
13. American Psychiatric Association, *American Psychiatric Association Practice Guidelines,* http://psychiatryonline.org/guidelines.aspx/ (April 22, 2012).
14. P. Bracken. "Psychiatry beyond the current paradigm." *Br J Psychiatry* 201 (2012): 430–4.
15. L. Cosgrove. "Conflicts of interest and disclosure in the American Psychiatric Association's clinical practice guidelines." *Psychother Psychosom* 78 (2009): 228–32.
16. L. Cosgrove. "The American Psychiatric Association's guideline for major depressive disorder: a commentary." *Psychother Psychosom* 81 (2012): 186–8. Also see, L. Cosgrove. "Conflicts of interest and the quality of recommendations in clinical guidelines." *J Eval Clin Practice* 19 (2013): 674–81.
17. American Psychiatric Association, *Practice Guideline for the Treatment of Patients with Major Depressive Disorder,* third edition (2012): 2–3.
18. The Carlat Psychiatry Blog. "A new smoking gun in the APA Textbook Fiasco," April 11, 2011.
19. Cosgrove, the APA's guideline for major depressive disorder: A commentary, ibid.
20. American Psychiatric Association, ibid, 3.
21. Cosgrove, Conflicts of interest in clinical guidelines, ibid.
22. American Psychiatric Association, ibid, 33.
23. House of Commons Health Committee, *The influence of the pharmaceutical industry* (London: House of Commons, 2005), http://www.publications.parliament.uk/pa/cm200405/cmselect/cmhealth/42/42.pdf/ (May 19, 2011).
24. B. Als-Nielsen. "Association of funding and conclusions in randomized drug trials: a reflection of treatment effect or adverse events?" *JAMA* 290 (2003): 921–8.
25. E. Turner. "Selective publication of antidepressant trials and its influence on apparent efficacy." *NEJM* 358 (2008): 252–60.
26. American Psychiatric Association, ibid, 32.
27. I. Kirsch. "Initial severity and antidepressant benefits." *PLoS Med* 5 (2008): 260–8.
28. J. Fournier. "Antidepressant drug effects and depression severity: a patient-level meta-analysis." *JAMA* 303 (2010): 47–53.
29. Fournier, ibid.
30. American Psychiatric Association, ibid, 31.
31. L. Bero. "Influences on the quality of published drug studies." *Int J of Technology Assessment in Health Care* 12 (1996): 209–37.
32. American Psychiatric Association, ibid, 82, 142.
33. A. Rush. "One-year clinical outcomes of depressed public sector outpatients." *Biol Psychiatry* 56 (2004): 46–53.
34. J. Rush. "Sequenced treatment alternatives to relieve depression (STAR*D): rationale and design." *Controlled Clinical Trials* 25 (2004): 119–42.

35. American Psychiatric Association, ibid, 98.

36. J. Van Scheyen. "Recurrent vital depressions." *Psychiatrica, Neurologia, Neurochirurgia* 76 (1973): 93–112.

37. Van Scheyen, ibid.

38. R. Hales, editor, *Textbook of Psychiatry* (Washington, DC: American Psychiatric Press, 1999): 525.

39. M. Shea. "Course of depressive symptoms over follow-up." *Arch Gen Psychiatry* 49 (1992): 782–7.

40. G. Fava. "Do antidepressant and antianxiety drugs increase chronicity in affective disorders?" *Psychother Psychosom* 61 (1994): 125–31.

41. G. Fava. "Holding on: depression, sensitization by antidepressant drugs, and the prodigal experts." *Psychother Psychosom* 64 (1995): 57–61.

42. R. Baldessarini. "Risks and implications of interrupting maintenance psychotropic drug therapy." *Psychother Psychosom* 63 (1995): 137–41.

43. W. Coryell. "Characteristics and significance of untreated major depressive disorder." *Am J Psychiatry* 152 (1995): 1124–9.

44. Fava, ibid.

45. American Psychiatric Association, ibid, 19.

46. El-Mallakh, R. "Tardive dysphoria: The role of long-term antidepressant use in inducing chronic depression." *Medical Hypotheses* 76 (2011): 769–73.

47. American Psychiatric Association, ibid, 50, 119.

48. J. Davidson. "Effect of Hypericum perforatum (St John's wort) in major depressive disorder." *JAMA* 287 (2002): 1807–14.

49. American Psychiatric Association, ibid, 29.

50. M. Babyak. "Exercise treatment for major depression." *Psychosomatic Med* 62 (2000): 633–8.

51. American Psychiatric Association, ibid, 17.

52. American Psychiatric Association, ibid, 17.

53. American Psychiatric Association, ibid, 17, 18.

54. American Psychiatric Association, ibid, 18.

55. American Psychiatric Association, ibid, 15–20.

56. American Psychiatric Association, ibid, 30.

57. American Psychiatric Association, ibid, 17.

58. National Institute for Health and Clinical Excellence. *Depression: the treatment and management of depression in adults. NICE clinical guideline 90,* (London: National Institute for Health and Clinical Excellence, 2009), http://www.nice.org.uk/nice-media/pdf/CG90NICEguideline.pdf/ (April 30, 2011).

59. National Collaborating Centre for Mental Health, *Depression: The NICE guidelines on the treatment and management of depression in adults* (updated edition). The British Psychological Society & The Royal College of Psychiatrists (2010): 484.

60. E. Van Weel-Baumgarten. "NHG standaard eepressie (tweede herziening)." *Huisarts en Wetenschap* 55 (2012): 252–9.

61. M. Murphy. "Alternative national guidelines for treating attention and depression problems in children." *Harv Rev Psychiatry* 22 (2014): 1–14.

62. S. Pliszka. "Practice parameter for the assessment and treatment of children and adolescents with attention-deficit/hyperactivity disorder." *J Am Acad Child Adolesc Psychiatry* 46 (2007): 894–921.

63. S. Pliszka. "The Texas Children's Medication Algorithm Project." *J Am Acad Child & Adolesc Psychiatry* 45 (2006): 642–57.

64. Murphy, ibid.

65. David J. Rothman, Expert witness report, October 15, 2010.
66. Rothman, ibid, 14–20.

9   A Society Harmed

1. D. Thompson. "Two concepts of corruption." Edmond J. Safra Working Papers, No. 16, August 1, 2013.
2. T. Beauchamp. "Informed consent: its history, meaning, and present challenges." *Cambridge Quarterly of Healthcare Ethics* 20 (2011): 515–23.
3. *Canterbury v. Spence*, 464 F.2d 772, 1972.
4. *Foy v. Greenblott,* 141 Cal. App. 3d. 190, 1983; *Mathis v. Morrissey*, 11 Cal. App. 4th 332 [13 Cal. Rptr. 2d 819], 1992.
5. Social Security Administration, annual statistical reports on the SSI and SSDI programs, 1987–2012. To calculate a total disability number for 1987 and 2012, we added the number of recipients under age 66 receiving an SSI payment that year and the number receiving an SSDI payment due to mental illness, and then adjusted the total to reflect the fact that one in every six SSDI recipients also receives an SSI payment. Thus, mathematically speaking: SSI recipients + (.833 × SSDI recipients) = total number of disabled mentally ill.
6. "Top therapeutic classes by U.S. sales, 2012." IMS Health.
7. C. Silverman. *The Epidemiology of Depression* (Baltimore, MD: Johns Hopkins Press, 1968): 139.
8. Social Security Administration, annual statistical reports on the SSDI and SSI programs, 2012.
9. R. Kessler. "Prevalence and treatment of mental disorders, 1990 to 2003." *NEJM* 352 (2005): 2515–23.
10. Social Security Administration, annual statistical reports on the SSDI and SSI programs, 1990, 2003.
11. Australian Government. "Characteristics of Disability Support Pension Recipients, June 2011."
12. New Zealand Ministry of Social Development. "National Benefits Factsheets," 2004–11.
13. S. Thorlacus. "Increased incidence of disability due to mental and behavioural disorders in Iceland, 1990–2007." *J Ment Health* 19 (2010): 176–83.
14. Danish government, The Appeals Board, Statistics on Early Retirement.
15. OECD Mental Health at Work: Sweden (2013).
16. Letter from the federal government to the minority members of Jutta Krellman, Sabine Zimmermann, Dr. Martina Bunge, and other Members of the Group of the Left, printed paper 17/9478, "Psychological stress in the workplace."
17. C. Miranda. "ADHD drugs could stunt growth." *Daily Telegraph* (UK), November 12, 2007.
18. Editorial. "Depressing research." *Lancet* 363 (2004): 1335.
19. E. Offidani. "Excessive mood elevation and behavioral activation with antidepressant treatment of juvenile depressive and anxiety disorders." *Psychother Psychosom* 82 (2013): 132–41.
20. G. Faedda. "Pediatric bipolar disorder." *Bipolar Disorders* 6 (2004): 305–13.
21. J. Kluger. "Young and Bipolar." *Time*, August 19, 2002.
22. M. Zoler. "U.S. children's psychiatric hospitalizations nearly doubled from 1996–2007." *Internal Medicine News,* November 19, 2010.

23. D. Satcher, *Report of Surgeon General's Conference on Children's Mental Health* (U.S. Dept. of Health and Human Services, 2001).

24. H. Marano. "Crisis on the campus." *Psychology Today*, May 2, 2002.

25. J. Blader. "Increased rates of bipolar disorder diagnoses among U.S. child, adolescent, and adult inpatients, 1996–2004." *Biol Psychiatry* 62 (2007): 107–14.

26. U.S. Government Accountability Office. "Young adults with serious mental illness: Some states and federal agencies are taking steps to address their transition challenges" (June 2008).

27. Social Security Bulletin, Annual Statistical Supplement, 1988; Social Security Administration, annual statistical report on the SSI program, 2011.

28. S. Kriss. "The Book of Lamentations." *TNI,* October 18, 2013.

29. Z. Hurston. *Dust Tracks on a Road: An Autobiography.* (New York: Harper Perennial Modern Classics, 2006): 227.

30. E-mail interview with Erin, July 15, 2014.

31. B. Pescosolido. "A disease like any other? A decade of change in public reactions to schizophrenia, depression, and alcohol dependence." *Am J Psychiatry* 167 (2010): 1321–30.

32. M. Harrow. "Factors involved in outcome and recovery in schizophrenia patients not on antipsychotic medications." *J Nerv Ment Disease* 195 (2007): 406–14. M. Harrow. "Do all schizophrenia patients need antipsychotic treatment continuously throughout their lifetime? A 20-year longitudinal study." *Psychol Med* 42 (2012): 2145–55.

33. B. Ho. "Long-term antipsychotic treatment and brain volumes." *Arch Gen Psychiatry* 68 (2011): 128–37. J. Radua. Also see B. Ho. "Progressive structural brain abnormalities and their relationship to clinical outcome." *Arch Gen Psychiatry* 60 (2003): 585–94. N. Andreasen. "Longitudinal changes in neurocognition during the first decade of schizophrenia illness." *International Congress on Schizophrenia Research* (2005): 348.

10   Putting Psychiatry on the Couch

1. S. Lewis. *I, Candidate for Governor: And How I Got Licked* (University of California Press, reprint edition, 1994): 109.

2. M. Banaji. "How (Un)ethical are you?" *Harv Business Rev* 81 (2003): 56–64.

3. H. Douglas. "Rejecting the Ideal of value-free science." Accessed at: http://uwaterloo. academia.edu/HeatherDouglas/Papers/1029159/Rejecting_the_Ideal_of_Value-Free_Science.

4. Young S. "Bias in the research literature and conflict of interest: an issue for publishers, editors, reviewers and authors, and it is not just about the money." *J Psychiatry Neurosci* 34 (2009): 412–17.

5. E. Harmon-Jones. "An introduction to cognitive dissonance theory and an overview of current perspectives on the theory," in *Cognitive Dissonance: Progress on a Pivotal Theory in Social Psychology.* (Washington DC: American Psychological Association, 1999).

6. C. Tavris. "Self-justification in public and private spheres." *The General Psychologist* 42 (2007): 4–7.

7. C. Tavris. "'Why won't they admit they're wrong?' and other skeptics' mysteries." *Skeptical Inquirer* 31 (2007): 12–13.

8. T. Wilson. "The unseen mind." *Science* 321 (2008): 1046–7.

9. Tavris, *The General Psychologist*, ibid.

10. Tavris, *Skeptical Inquirer* 31, ibid.

11. M. Morgan. "Interactions of doctors with the pharmaceutical industry." *J Med Ethics* 32 (2006): 559–63.
12. M. Steinman. "Of principles and pens." *Am J Med* 110 (2001): 551–7.
13. B. Hodges. "Interactions with the pharmaceutical industry." *CMAJ* 153 (1995): 553–9.
14. N. Choudhry. "Relationships between authors of clinical practice guidelines and the pharmaceutical industry." *JAMA* 287 (2002): 612–17.
15. S. Chimonas. "Physicians and drug representatives: exploring the dynamics of the relationship." *J Gen Intern Med* 22 (2007): 184–90.
16. S. Sah. "Effect of reminders of personal sacrifice and suggested rationalizations on residents' self-reported willingness to accept gifts." *JAMA* 304 (2010): 1204–11.
17. P. Appelbaum. "Psychiatrists' relationships with industry: the principal-agent problem." *Harv Rev Psychiatry* 18 (2010): 255–65.
18. A. Detsky. "Sources of bias for authors of clinical practice guidelines." *CMAJ* 175 (2006): 1033.
19. G. Alexander. "Brief report: physician narcissism, ego threats, and confidence in the face of uncertainty." *J Applied Social Psychology* 40 (2010): 947–55.
20. G. Greenberg. "Inside the battle to define mental illness." *Wired Magazine*, December 27, 2010.
21. G. Klerman. "A debate on DSM-III." *Am J Psychiatry* 141 (1984): 539–42.
22. J. Talbott. "Response to the presidential address: psychiatry's unfinished business in the 20th century." *Am J Psychiatry* 141 (1984): 927–30.
23. C. Nadelson. "Response to the presidential address: unity amidst diversity." *Am J Psychiatry* 142 (1985): 1009–14.
24. R. Pasnau. "Presidential address: psychiatry in medicine.'" *Am J Psychiatry* 144 (1987): 975–80.
25. P. Fink. "Response to the presidential address: is 'biopsychosocial' the psychiatric shibboleth?" *Am J Psychiatry* 145 (1988): 1061–7.
26. H. Pardes. "Presidential address: defending humanistic values." *Am J Psychiatry* 147 (1990): 1113–19.
27. J. English. "Presidential address: patient care for the twenty-first century." *Am J Psychiatry* 150 (1993): 1298–301.
28. H. Sacks. "Response to the presidential address: new challenges for proven values." *Am J Psychiatry* 154 (1997): 1350–3.
29. S. Sharfstein. "Presidential address: advocacy as leadership." *Am J Psychiatry* 163 (2006): 1711–15.
30. N. Stotland. "Presidential address." *Am J Psychiatry* 166 (2009): 1100–4.
31. A. Schatzberg. "Presidential address." *Am J Psychiatry* 167 (2010): 1161–5.
32. C. Bernstein. "Response to the presidential address." *Am J Psychiatry* 167 (2010): 1166–9.
33. J. Lieberman. "Presidential address: our future is now." *Am J Psychiatry* 171 (2014): 733–7.
34. Alexander, ibid.
35. A. Nierenberg. "A counter proposal to manage financial conflicts of interest in academic psychiatry." *World Psychiatry* 6 (2007): 34–6.
36. M. Thase. "On the propriety of collaborations between academicians and the pharmaceutical industry." *World Psychiatry* 6 (2007): 29–31.
37. Nierenberg, ibid.
38. M. Elias. "Conflicts of Interest Bedevil Psychiatric Drug Research." *USA Today*, June 3, 2009.

39. L. Cosgrove. "A comparison of DSM-IV and DSM-5 panel members' financial associations with industry: a pernicious problem persists." *PLoS Med* 9 (2012): e1001190.

40. D. Brauser. "APA answers criticism of pharma-influenced bias in DSM-5." *Medscape*, January 4, 2013.

41. J. Lieberman. "Time to Re-engage with Pharma?" *Psychiatric News* 48, August 1, 2013.

42. Lieberman, ibid.

43. R. Pies. "Psychiatry's New Brain-mind and the Legend of the 'Chemical Imbalance.'" *Psychiatric Times*, July 11, 2011.

44. R. Pies. "Doctor, is my mood disorder due to a chemical imbalance?" Psychcentral. com, August 4, 2011.

45. NPR, Fresh Air show: "A psychiatrist's prescription for his profession." July 13, 2010.

46. NPR, Morning Edition. "When it comes to depression, serotonin isn't the whole story." January 23, 2012.

47. R. Harding. "Unlocking the brain's secrets." *Family Circle*, November 20, 2001, 62.

48. American Psychiatric Association. www.psychiatry.org/mental-health/depression. Accessed on July 23, 2014.

49. The Balanced Mind Parent Network. "Fact Sheet: Facts About Teenage Depression," November 27, 2009. Accessed June 29, 2014, at www.thebalancedmind.org/learn/ library/facts-about-teenage-depression.

50. Depression and Bipolar Support Alliance. "Coping with mood changes later in life." Accessed June 29, 2014 at www.dbsalliance.org/site/ PageServer?pagename=education_brochures_coping_mood_changes

51. NAMI. "Major Depression Fact Sheet." Accessed June 29, 2014 at www.nami. org/Content/Microsites270/NAMI_Howard_County/Home258/Mental_Illness_ Information1/Depression.pdf

52. B. Pescosolido. "A disease like any other?" *Am J Psychiatry* 167 (2010): 1321–30; see table 1, p. 1324.

53. J. Kemp. "Effects of a chemical imbalance causal explanation on individuals' perceptions of their depressive symptoms." *J Behav Res Therapy* 56 (2014): 47–52.

54. Fournier J. "Antidepressant drug effects and depression severity." *JAMA* 303 (2010): 47–53.

55. P. Kramer. "In Defense of Antidepressants." *New York Times*, July 9, 2011.

56. American Psychiatric Association, News release: "'60 Minutes' segment on antidepressants 'irresponsible and dangerous.'" February 22, 2012.

57. M. Angell. "The Illusions of Psychiatry." *The New York Review of Books*, July 14, 2011.

58. M. Moran. "Prominent M.D.'s Book Reviews Give Negative View of Psychiatry." *Psychiatric News* 46, August 5, 2011, 8

59. J. Oldham, Letter to the Editor, *The New York Review of Books*, August 18, 2011.

60. D. Regier. "Book Review Misses Evolution of Psychiatry." Issued as an "official APA response," August 5, 2011. See excerpt in *Psychiatric News*, August 5, 2011.

61. R. Friedman and A. Nierenberg, Letters to the Editor, *The New York Review of Books*, August 18, 2011.

62. A. Frances. "A Warning Sign on the Road to DSM-V: Beware of Its Unintended Consequences." *Psychiatric Times*, June 26, 2009.

63. A. Frances and R. Spitzer. "Letter to the APA Board of Trustees," July 6, 2009.

64. A. Schatzberg. "Setting the Record Straight: A Response to Frances Commentary on DSM-V." *Psychiatric Times,* July 1, 2009.
65. J. Lieberman. "DSM-5: Caught between mental illness stigma and anti-psychiatry prejudice." Guest blog on *Scientific American*, May 20, 2013.

11 Prescriptions for Reform

1. D. Davis. *The Secret History of the War on Cancer* (Basic Books, reprint edition, 2009): 435.
2. B. Falit. "Curbing industry sponsor's incentives to design post-approval trials that are suboptimal for informing prescribers but more likely than optimal designs to yield favorable results." *Seton Hall Law Review* 37 (2007): 969–1049. See page 971.
3. Falit, ibid.
4. M. Bazerman and Ann E. Tenbrunsel. *Blind Spots: Why We Fail to Do What'ss Right and What to Do about It* (Princeton, NJ: Princeton University Press, 2011).
5. T. Stossel. "Opinion: What's Wrong with COI?" *The Scientist*, June 12, 2012.
6. See T. Mendelson. "Conflicts of interest in cardiovascular clinical practice guidelines." *Arch Intern Med* 171 (2011): 577–84. This study shows that this type of excuse, that there are few nonconflicted experts, is not true for cardiology. While financial ties to industry are undoubtedly more prevalent in psychiatry, there are an increasing number of academic psychiatrists today without such conflicts.
7. S. Sah. "Physicians under the influence." *J Law Med Ethics* 41 (2013): 665–72.
8. L. Cosgrove. "Pharmaceutical Philanthropic Shell Games." *Psychiatric Times*, March 6, 2010.
9. C. Robertson. "Blinding as a solution to institutional corruption." Edmond J. Safra working paper 21, September 5, 2013.
10. P. Bracken. "Postpsychiatry: a new direction for mental health." *Br Med J* 322 (2001): 724–7.

INDEX

Page numbers in italics refer to content in tables and figures.